Thaís Rosa
Diogo Helgueira
Luis Antonio de Avila

Seed treatment, herbicides, air temperature in irrigated rice

Thaís Rosa
Diogo Helgueira
Luis Antonio de Avila

Seed treatment, herbicides, air temperature in irrigated rice

Stresses under irrigated rice cultivation

Imprint

Any brand names and product names mentioned in this book are subject to trademark, brand or patent protection and are trademarks or registered trademarks of their respective holders. The use of brand names, product names, common names, trade names, product descriptions etc. even without a particular marking in this work is in no way to be construed to mean that such names may be regarded as unrestricted in respect of trademark and brand protection legislation and could thus be used by anyone.

Cover image: www.ingimage.com

This book is a translation from the original published under ISBN 978-3-330-75879-7.

Publisher:
Sciencia Scripts
is a trademark of
Dodo Books Indian Ocean Ltd. and OmniScriptum S.R.L publishing group

120 High Road, East Finchley, London, N2 9ED, United Kingdom
Str. Armeneasca 28/1, office 1, Chisinau MD-2012, Republic of Moldova, Europe
Printed at: see last page
ISBN: 978-620-3-63116-6

Summary

Effect of seed treatment on the initial establishment of irrigated rice at different temperatures and sowing times

ABSTRACT - The aim of this study was to evaluate the effects of seed treatment on the initial establishment of irrigated rice plants and their tolerance to herbicides at low temperatures. Two experiments were carried out, one in the laboratory and the other in the field. In the first experiment, the experimental design used was entirely randomized, in a factorial scheme (7x2) where factor A consisted of six seed treatments T1: control; T2: thiamethoxam; T3: dietholate; T4: fipronil; T5: GA3; T6: carboxin+thiram; T7: fipronil+dietholate+carboxin+thiram, and factor B consisted of two temperatures 25 and 17°C, with the following evaluations: germination test, first germination count, cold test, accelerated ageing, germination speed index, shoot and root length, dry matter mass. In the field experiment, a randomized block design was used in a factorial scheme (7x3x2) where factor A consisted of the same seed treatments, factor B consisted of two herbicides bispiribaque-sodium (50 g i.a ha^{-1}) and profoxidim (170 g i.a ha^{-1}) and a control with no herbicide and factor C with two sowing times (September and October). The evaluations were phytotoxicity at seven, 14, 21 and 28 days after application of the herbicides, initial stand, number of tillers, number of panicles, grains/panicle, weight of 100 grains and yield. The laboratory results show that gibberellic acid provides greater initial germination of rice seeds under cold stress and at ideal temperatures, with the treatment with dietholate and the combination dietholate+fipronil+carboxin+thiram showing decreases in the variables evaluated at both temperatures 25 and 17°C. In the field, there was no effect of seed treatment and there was greater phytotoxicity of the herbicide profoxidim in rice plants sown in October, but this factor did not influence their productivity, obtaining higher yields than rice sown in September.

Keywords: initial performance, abiotic stress, Oryza sativa, selectivity.

INTRODUCTION

Rio Grande do Sul stands out as the main rice producing state in Brazil, growing irrigated rice on an area of 1.12 million hectares, with an average productivity of 7.2 t.ha^{-1} (CONAB, 2014). Rice cultivation in Rio Grande do Sul is of fundamental importance in the state's agro-economic scenario, accounting for around 61% of the total volume of rice produced in Brazil (IRGA, 2012). However, the crop is subject to stresses that can compromise its productive potential, including biotic factors such as pests and weeds and abiotic factors such as temperature and light.

To this end, alternatives have been sought to make better use of the genetic potential of cultivars and reduce production costs. One of the alternatives that has stood out in recent years is sowing at the beginning of the preferential season, providing gains in yield, combined with the search for alternatives that allow the treated plants to be more tolerant of stress factors and consequently to develop vigorously in sub-optimal conditions, allowing better chances of achieving higher yields (Almeida et al., 2013).

In RS, early sowing, carried out in late September and early October, synchronizes the period of greatest availability of solar radiation, which occurs in December and January in Rio Grande do Sul (Steinmetz et al., 2001), with the reproductive phase of rice, an important factor for obtaining high yields due to the greater efficiency in nitrogen assimilation during the stage of microsporogenesis (pollen grain formation) and grain filling (Freitas et al., 2008).

On the other hand, one of the problems caused by early sowing is low temperature stress, which can result in low seed germination and, consequently, low initial seedling stand, resulting in yield losses and reduced tolerance to herbicides. This report has been observed mainly in the Campanha and Sul do RS regions, where early sown crops show high phytotoxicity caused by herbicides classified as selective (Martini, 2014).

Herbicide selectivity is a key tool in weed control and depends on the interaction of three main factors: the crop, the herbicide and the environment. Low temperatures can influence the increase in the concentration of reactive oxygen species (ROS), as they generate negative effects on plant metabolism (Hashimoto & Komatsu, 2007). The metabolism of herbicides can also be directly affected by this increase (Vila-Aiub et al., 2012), which is closely related to a reduction in herbicide tolerance and can cause a decrease in productivity.

Another factor that can limit the productivity of irrigated rice in RS is competition with weeds (Lilge et al., 2003). More effective control of weeds by herbicides without causing severe damage to the crop is desirable. To this end, one of the alternatives used is seed treatment with protectants to minimize the phytotoxic effects caused by some herbicides.

Some products have a protective action and this is related to increased plant tolerance, allowing them to metabolize the chemical faster than weeds, causing less damage to the crop (Yazbek Júnior and Foloni, 2004).

Seed treatment is a practice used to minimize productivity losses caused by pests and diseases that can affect the initial development of crops. This practice is based on applying chemical products such as insecticides, fungicides, growth regulators or even micronutrients to the seeds in the period before sowing. In addition to the protection that the product offers the seeds, studies indicate that there may be an improvement in the speed of seedling emergence (Almeida et al., 2014).

Many studies attribute some products used for seed treatment to a certain phytotonic effect, i.e. faster seedling development, better germination and greater vigor (Almeida et al., 2014). This effect may be relevant, since factors such as temperature directly regulate the speed of this process (Marcos Filho, 2005).

The aim of this study was to evaluate the effects of seed treatment on the initial establishment of irrigated rice plants and their tolerance to herbicides when subjected to low temperatures.

MATERIAL AND METHODS

The study comprised two experiments, the first of which was carried out in the Seeds and Technology Laboratory of the Department of Plant Science at the Federal University of Pelotas (UFPel) and the second of which was conducted in the field, in a systematized area, at the Centro Agropecuàrio da Palma (CAP), belonging to the Federal University of Pelotas (UFPel), during the 2013/14 agricultural year. The soil at the site is classified as Planossolo Hàplico Eutròfico solódico and belongs to the Pelotas mapping unit (EMBRAPA, 2012). The physico-chemical characteristics of the 0-0.2 m deep layer of soil in which the experiment was conducted are shown in Table 1:

Table 1. Physico-chemical characteristics of the 0-0.2 m depth layer of the

soil (Planossolo Hidromorfico Eutròfico solódico) of the experimental area. FAEM/UFPel - Capao do Leao, RS, 2014.

pHWater	Clay	M.O	Phosphorus	Potassium	Calcium	Magnesium	Aluminum
(1:1)	(%)		mg	dm^{-3}	cmolc dm^{-3}		
5, 7	161	,3	6,7	43	4,	32,0	0,1

Experiment 1 - Effect of different seed treatments on the physiological potential of irrigated rice seeds

The experiment was arranged in a 7x2 factorial design, in a completely randomized design, with four replications. Factor A consisted of a control, without seed treatment, and six products (or combinations of products) for seed treatment (Table 2). Factor B consisted of two environmental temperatures (25 and 17°C) for the development of rice seedlings. The IRGA 424 rice cultivar was used, chosen due to its high production potential and good adaptation to low temperature conditions (SOSBAI, 2012).

Table 2. Products used for seed treatment (ST) in irrigated rice cultivation. FAEM/UFPel - Capao do Leao, RS, 2013/14.

Treatments	Factor A: Active Ingredient in TS	Dose g a.i. 100 kg^{-1}

1	No application	---
2	Thiamethoxam	140,0
3	Dietholate	600,0
4	Fipronil	62,5
5	Gibberellic acid	2,0
6	Carboxin+Tiram	60,0 + 60,0
7	Dietolate+Fipronil+Carboxine+Tiram	600,0+62,5+60,0+60,0

The seed treatments were carried out directly on the seeds using a pressurized valve, 24 hours before the experiments were set up. The seeds were placed in plastic bags with a capacity of five liters, using one (1) kg of seeds per bag. The volume of spray used was 1.5 L 100 kg^{-1} of seeds and, for the control treatment, only distilled water was used.

The influence of seed treatment and environmental temperatures on the different physiological characteristics of rice seeds was evaluated using the following analyses:

1)   Germination test: carried out on four sub-samples of 50 seeds, totaling 16 experimental units per treatment, arranged to germinate in rolls made up of three sheets of germitest paper, moistened with distilled water 2.5 times the weight of the dry paper. The rolls were transferred to a BOD germination chamber at 17 °C and 25 °C with a 12h photoperiod. Evaluations were carried out 14 days after sowing and the results expressed as a percentage of normal seedlings, according to the Seed Analysis Rules (Brasil, 2009).

2)   First germination count: conducted in conjunction with the germination test, five days after sowing, in accordance with the RAS (Rules for Seed Analysis), for temperatures of 25 and 17°C. The results were expressed as a percentage of normal seedlings.

3)   Germination speed index (GVI): obtained from daily counts of germinated seeds (minimum root protrusion of 3 to 4 mm). The counts were carried out until a constant number of germinated seeds was obtained. The equation suggested by Popinigis (1985) was used to calculate the germination speed index (GVI):

IVG=N1/D1+N2/D2+Nn/Dn, where:

N_i= number of seedlings emerged on the first day;

N_n= accumulated number of emerged seedlings;

D_i= first day of counting;

D_n= number of days after sowing.

4)   Accelerated **ageing**: The seeds were placed in acrylic boxes (mini-chambers) of the gerbox type with a horizontal metal screen fixed in the middle position. Fifty seeds for each treatment were evenly distributed on the screen and 40 ml of distilled water was added (to obtain approximately 100% R.U.). The boxes were then closed and placed in a BOD incubator with the temperature set at 41°C, where they remained for 120 hours (AOSA, 1983). After this ageing period, the seeds were placed on germitest paper and allowed to germinate under the same conditions as the germination test at temperatures of 17 and 25°C. They were evaluated on the fifth day after the test. The results were expressed as a percentage of normal seedlings.

5)   **Cold test**: carried out on four subsamples of 50 seeds per sample, distributed on a roll of germ paper previously moistened with distilled water at a rate of 2.5 times the weight of the dry paper and subjected to constant temperature. The rolls were covered with plastic bags to prevent moisture loss and kept in a B.O.D. at a temperature of 10°C for a period of seven days, according to the methodology proposed by the Vigor Committee of the International Seed Testing Association (ISTA, 1995). After this period, the rolls were transferred to a germinator at 25 and 17°C, and the results expressed as a percentage of normal seedlings, in accordance with the Seed Analysis Rules (Brasil, 2009).

6)   **Length of aerial part and root**: four replicates of 10 seeds were used, sown on germitest paper. After seven days, the root and aerial part **lengths** of the normal seedlings were assessed using a millimeter ruler.

7)   **Seedling dry matter mass**: was determined on four replicates of ten seedlings after 14 days, and kept in paper bags in an oven at 60°C until a constant mass was obtained, weighed on a precision balance (0.001 g). The value obtained from the sum of each repetition was divided by the number of seedlings used and the results were expressed in mg. seedling $.^{-1}$

The data was subjected to analysis of variance and, when the effect of the treatments was proven to be significant by the F test ($p \leq 0.05$), the means of the treatments were compared using the Tukey test ($p \leq 0.05$). The data for the germination percentage characteristic was transformed according to RAIZ $(Y+1)1$

Experiment 2 - Effect of seed treatment on the initial establishment of irrigated rice at different sowing times in the field

The study was conducted in the field during the 2013/14 agricultural year, in a systematized area, at the Centro Agropecuàrio da Palma (CAP), belonging to the Universidade Federal de Pelotas (UFPel). The experiment was arranged in a 7x3x2 factorial scheme, in a randomized block experimental design with four replications.

Factor A consisted of six products used in seed treatment for irrigated rice, at the doses recommended by the manufacturer, and a control (no seed treatment), as shown in Table 2.

Factor B consisted of two herbicide treatments and a control (Table 3).

Table 3. Herbicide treatments applied to irrigated rice for weed control. FAEM/UFPel - Capao do Leao, RS, 2013/14.

Factor B:	Registration dose of	
Post-emergence Active Ingredient p.c	(g i.a. ha $)^{-1}$	Application Season
control	-	-
bispiribaque sodium	50	Post-emergence
profoxidim	170	Post-emergence

Factor C corresponds to two sowing seasons which were defined based on the agro-climatic zoning for rice cultivation in the state of Rio Grande do Sul (SOSBAI, 2012). The first sowing season was implemented on September 13th (before the recommended period) and the second on October 18th (recommended period).

The experimental units consisted of nine sowing lines spaced 17 cm apart and five meters long, totaling 7.65 m^2 . The soil was prepared using the conventional cultivation system. The rice cultivar used was IRGA 424, at a density of 100 kg of seeds ha .$^{-1}$

The base fertilizer was 350 kg ha^{-1} of the 05-20-30 formula, corresponding to a supply of 17.5 kg ha^{-1} of nitrogen (N), 70 kg ha^{-1} of P O_{25} and 105 kg ha^{-1} of K_2 O. Top dressing nitrogen fertilization was carried out with urea (46-00-00) in three stages. The first application was made at the beginning of tillering (V -V_{34}), with 70 kg ha^{-1} of N applied, the second application was made during the V_6 stage, with 30 kg ha^{-1} of N applied and the third application was made at the flower bud stage, with 30 kg ha^{-1} of N applied. The other crop treatments were carried out in accordance with the technical indications of the research for growing irrigated rice in southern Brazil (SOSBAI, 2012).

The experimental area had a natural infestation of rice grass (*Echinochloa* sp.), and the average population in the experiment was 500 plants m^{-2} . Before the installation of each sowing season, the area was desiccated with the application of the herbicide glyphosate at a dose of 1440 g i.a. ha^{-1} . The herbicide treatments were applied post-emergence to irrigated rice, when the weeds were in the physiological stage V -V_{34} (3 to 4 leaves), following the manufacturer's recommendations for use.

The application was carried out using a precision knapsack sprayer, pressurized with CO_2 , equipped with a boom with four flat fan nozzles, 110-02 series, spaced 0.5 m apart, calibrated to apply a spray volume of 150 l ha^{-1} . Shortly after the herbicides were applied, the first top dressing of nitrogen was applied. The day after the application, the area was flooded, keeping the water line at eight cm until the crop was physiologically ripe.

The evaluations carried out were: phytotoxicity, initial stand, number of tillers, number of panicles, number of grains per panicle, weight of 100 grains and yield.

Phytotoxicity was assessed visually at seven, 14, 21 and 28 days after herbicide application (DAH) by assigning scores based on a percentage scale from 0 to 100%, where zero corresponds to the absence of injury and 100% corresponds to plant death (Sociedade Brasileira da Ciência das Plantas Daninhas 1995).

The initial stand was determined at 10 days after emergence (DAE) by counting the number of plants in one meter of the sowing line. At this point, the number of tillers and the number of panicles per square meter were determined, and by collecting 10 panicles in sequence from the sowing line, the number of grains per panicle and the mass of one hundred grains were determined.

In order to assess grain yield, the useful area of each plot (4.76 m^2) was harvested by hand when the grains had reached an average moisture content of 22%. This material was sorted, weighed and the grain's harvest moisture determined, the latter being corrected to 13% in order to estimate yield.

The data obtained was previously analyzed for compliance with the assumptions of analysis of variance (normality and homoscedasticity of variance $p \leq 0.05$), transforming them when necessary. If the effects of the treatments were significant, the Tukey test ($p \leq 0.05$) was applied to compare the means of the herbicide treatments and the t-test ($p \leq 0.05$) was applied to compare the sowing times.

RESULTS AND DISCUSSION

Experiment 1-   Effect of different seed treatments on the potential physiological potential of irrigated rice seeds

There was an interaction between the factors seed treatment and temperature. Under the conditions that the rice seeds were subjected to, it was observed that seed treatment influenced the initial development of seedlings at temperatures of 25 and 17°C.

With regard to germination, there was a significant difference between the temperatures and, when the seeds were subjected to the optimum temperature (25°C), the differences between the products and the control (untreated) became less significant, with only the treatments with dietholate and the combination of dietholate+fipronil+carboxin+t differing from the others, resulting in the lowest germination values (Table 4).

When subjected to cold stress (17°C), the effect of the seed treatment became more visible, with a difference between the treatments (Table 4).

The treatment containing GA3 stood out in terms of seed germination when it was carried out at a temperature of 17°C. This is possibly due to the effect of the hormone, considered an endogenous enzyme activator (Levitt, 1974), on germination uniformity.

Há several reports confirming the improvement in germination through the use of GA3, such as the result in the poaceae Trisacum dactyloides, observed by Rogis et al. (2004). The use of growth regulators such as gibberellins (Bevilaqua et al., 1993), cytokinins (Cunha & Casali, 1989) and etrel (Suge, 1971) during the germination phase can improve the performance of seeds of various species, especially under adverse conditions. This factor was observed in the results obtained when the seeds were under low temperature stress (17°C). However, the influence of GA3 on seed germination depends on the species (King et al., 1987).

In the germination assessment at both temperatures, the application of the insecticide thiamethoxam showed no difference compared to the treatment without seed treatment (control). According to the results presented by Almeida (2011), in controlled conditions, in which different rice cultivars were used, including the IRGA 424 cultivar, and the physiological performance of rice seeds was evaluated with the application of thiamethoxan, the results showed an improvement in the physiological quality of the seeds.

It was observed that the use of dietholate negatively affected germination and the first germination count at 25°C (Figure 3) and 17°C, when used alone or in combination with insecticide and fungicide (Table 4). This combination is commonly used by most producers, but in the germination test, this was the treatment that most damaged the germination process in both temperature conditions.

Table 4 Germination, first germination count, germination speed index, cold test for rice seeds treated with different products and subjected to different temperatures.

Treatments	Germination (%) '		First Germination Count (%)		Germination Speed Index (JVG) (%)		Cold test (%)	
	25°C	17°C	25°C	17°C ¡ 25°C		17°C	25°C	17°C
Tl-without treatment	91,4 to[1] A[2]	43,0 cB	96,1 ab[1] A[2]	16,1 bcdB	21,7 d A[12] 3,1	aB	92,0 to[1] A[2]	27,1 cB
T2-thiamethoxam	90,6 aA	35,8 cB	95,0 aA	13,0 cdB	47,1 abA 2,0	abB	77,1 cdA	35,0 cB
T3-dietholate	86,6 bA	37,1 cB	85,0 bA	9,3 deB	42,2 cA 1,2	bB	75,2 cdA	33,1 cB

Treatment																
T4-fipronil	90,8	aA	46,2	bcB	96,0	aA	17,2	bcB	48,1	aA	2,4	abB	82,3	bcA	37,8	bcB
T5-GA$_3$	92,2	aA	78,3	aB	95,2	aA	36,1	aB	47,8	abA	3,8	aB	90,0	abA	**64,1**	aB
T6-carboxin + thiram	94,4	aA	62,4	bB	93,0	aA	22,0	bB	46,1	bA	3,4	aB	89,0	abA	37,7	bcB
T7-DFCT[1]	85,1	cA	15,3	dB	82,2	bA	4,2	eB	41,2	cA	0,9	bB	67,3	dA	18,2	dB
CV (%)		15,5				15,03				3,97				9,2		

[1] Averages with different capital letters in the row differ by student's t-test (p≤0.05).

[2] Averages with different lower-case letters in the column differ by the Tukey test (p≤0.05)J

[3] Dietolate, Carboxin + thiram - fipronil combinations.

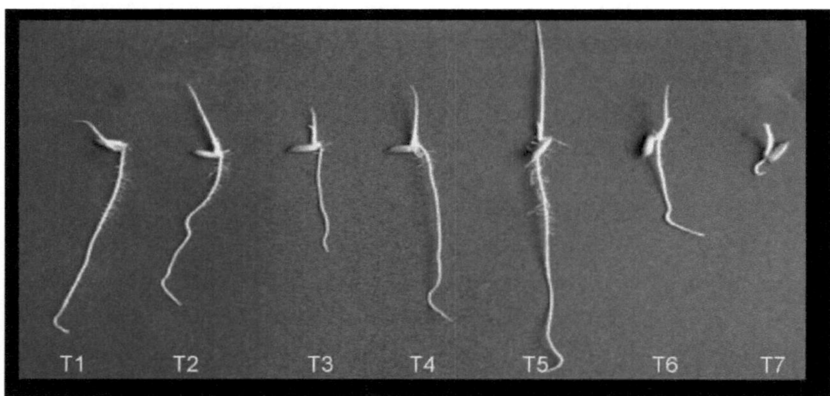

Figura 1. Visualization of the effect of seed treatments on the IRGA 424 cultivar in the first germination count at 25 °C. T1: no application; T2: thiamethoxam; T3: dietholate; T4: fipronil; T5: GA3; T6: carboxin + thiram; T7: dietholate + fipronil + carboxin + thiram.

According to Deridder et al. (2002), seed protectants increase the tolerance of cereals to chemical products through physiological mechanisms. The protective effect of dietholate is related to increased expression of the enzyme glutathione S-transferase (GSTs), causing changes in the plant by activating oxidation, reduction and hydrolysis reactions. Thus, the reduction in germination was possibly due to changes in the metabolism of the treated seeds during the germination phase.

For the treatment consisting of a combination of the products dietholate+fipronil+carboxin+thiram, there was a reduction in the percentage of germination conducted at both low temperature (17°C) and optimum temperature (25°C) (Table 4). Possibly, there was an interaction between the products applied, resulting in a reduction in normal seedlings, combined with low temperature stress (Mertz, et al., 2009).

Gibberellic acid was the treatment that most affected vigor, as represented by the first germination count at five days after sowing (five DAS), especially when subjected to a temperature of 17°C (Table 4). The stimulation of germination with gibberellic acid is due to the rapid synthesis of hydrolytic enzymes in response to the presence of the hormone (GA3). These enzymes are transported to the aleurone layer of the seed, which is responsible for converting starch into sugar and is therefore used in the growth of the new seedling (Schwechheimer, 2008).

According to the results obtained, there was an improvement in the initial performance of rice seeds with the application of GA3, represented by the evaluation of the first germination count. The use of a growth regulator for seed treatment can positively influence the development of rice seedlings, helping with the initial establishment of seedlings subjected to stress (Bevilaqua et al., 1993).

Despite being considered a cultivar with good adaptation to medium and low temperature conditions in the cold sensitivity test, the IRGA 424 cultivar showed survival of only 2.5% of seedlings in the work carried out by Cruz et al. (2010).

It should be noted that when germination is high, these results will not guarantee a similar performance later on, as this depends on the physiological potential and environmental conditions. According to Krohn & Malavasi (2004), treated seeds outperform those that have not received any treatment in most cases.

In the evaluation of the first germination count at 25°C, the thiamethoxam, fipronil and carboxin+thiram treatments did not differ from each other. On the other hand, the treatments with dietholate and the combination of dietholate+fipronil+carboxin+thiram reduced the germination percentage, due to the degree of sensitivity of the treated seeds, which contributed to the loss of vigor (Table 4).

The same occurred at 17°C, when the thiamethoxam, fipronil, carboxin+thiram and control treatments did not differ from each other. However, there was a significant decrease in normal seedlings in this evaluation for the dietholate and dietholate+fipronil+carboxin+thiram treatments (Table 4).

One of the techniques used to estimate seed vigor is the germination speed index, whereby the faster the seed germinates, the greater its vigor (Lima et al., 2005). The results obtained in this study allow us to infer that applying treatments to seeds increases the initial establishment of rice seedlings subjected to a temperature of 25°C, compared to untreated seeds (Table 4). The treatments thiamethoxam, fipronil, GA3 and carboxin+thiram did not differ from each other.

Seeds treated with dietholate and the combination of dietholate+fipronil+carboxin+thiram showed a reduction in vigor. This is possibly because the seeds have some degree of sensitivity to the application of the products, which contributes to the reduction in vigor. In addition, another hypothesis is that covering the seeds with this protectant reduces the speed of water absorption, which is essential for triggering the metabolic and biochemical processes of germination, thus causing a reduction in GVI.

The control treatments (no treatment), thiamethoxam, dietholate, fipronil and carboxin+thiram, and no treatment showed no significant difference at 17°C (Table 4).

According to Almeida et al. (2009), thiamethoxam, among the treatments tested, stimulated the physiological performance of carrot seeds subjected or not to water stress, but in other stress conditions (temperature), these results were not observed.

The cold test is based on assessing seed quality under adverse temperature conditions and is commonly used to differentiate levels of vigor related to seed shape, weight, size and treatment (Marcos-Filho et al., 1977 and Silva & Marcos-Filho, 1979).

Table 4 shows the results of the cold test, where the control treatment did not differ from the treatments with GA3 and carboxin + thiram at a temperature of 25°C, showing similar results in the percentage of normal seedlings. The thiamethoxam, dietholate and fipronil treatments did not differ. The combination of the products dietholate + fipronil + carboxin + thiram reduced the number of normal seedlings in the test, differing from the other treatments, and was therefore the treatment that most influenced the vigor of the rice seedlings.

For the combination of dietholate + fipronil + carboxin + thiram, there was a decrease in vigor for the cold test at 17°C (Table 4). This can possibly be explained by the interaction between the products, and another possible hypothesis is that the seed first metabolizes the product, wasting energy and causing an energy deficit for growth.

(Salgado et al., 2013), coupled with the temperature conditions under which the seeds are subjected to the cold test.

The accelerated ageing test consists of exposing seeds to high humidity and temperature conditions, a process which causes deterioration and reduces the sugar content of the seeds (Kapoor et al., 2011).

When evaluating the effect of seed treatment in the accelerated ageing test, the results did not differ from the other results observed, with the negative effect standing out when the seeds received the product dietholate, which was evident at a temperature of 25°C, and especially when combining the products dietholate+fipronil+carboxin+thiram at a temperature of 17°C (Table 5).

When the seeds were exposed to stress conditions (low temperature), after undergoing a deterioration process as occurs with the accelerated ageing test, the treatment with gibberellic acid proved to be effective, significantly increasing the percentage of normal seedlings (Figure 2).

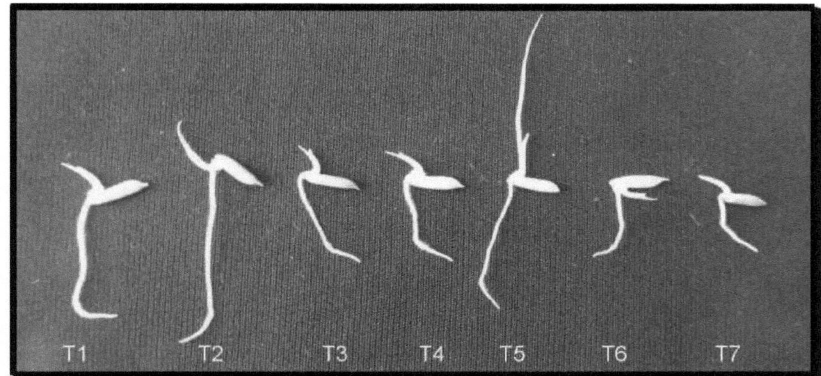

Figura 2. Accelerated aging test of cultivar IRGA 424. T1: no application; T2: thiamethoxam; T3: dietholate; T4: fipronil; T5: GA3; T6: carboxin + thiram; T7: dietholate + fipronil + carboxin + thiram.

There was an increase in the length of the aerial part in seeds treated with gibberellic acid (Table 5). These results were also evidenced in a study with different rice cultivars by Flores et al. (2002), in which the action of GA3 at the seedling stage found significant differences in seedling length.

Other authors have also observed similar responses to the *action* of gibberellic acid (Dias et al. 1995, Peske et al., 1991, Souza and Menezes, 1991 and Bevilaqua et al., 1993).

The reduction in growth suggests that the rice seeds are sensitive to the protectant studied, since the treatments containing dietholate inhibited the initial growth of the aerial part of the seedlings at both temperatures tested (Table 5). Similar results were obtained in sorghum, using the seed protectant flurazole, which led to inhibition of seedling growth (Hirase & Molin, 2003), and in rice, with the protectant diethylphenylphosphorothioate (Mistura, 2008).

Table 6 shows the results for the root length of the rice seedlings. There was no difference between the treatments at 25°C. This result may be related to the fact that root development is more dependent on temperature than the aerial part (Mendany et al., 2007).

When subjected to stress conditions, there were differences between the treatments. The control, thiamethoxam, fipronil, GA3 and carboxin+thiram treatments did not differ (Table 5). Treatment with the combination of dietolate+fipronil+carboxin+thiram reduced root length, but did not differ from the control, fipronil and GA3 treatments (Table 5).

Although the use of gibberellic acid positively influenced the other tests, it did not affect root length. Studies have shown that the growth regulator has no effect on the root system because it develops the growth of the aerial part (Wahyuni et al. 2003), which was evident in this study at a temperature of 25°C. However, Flores et al. (2002) observed a positive effect of GA3 on root length in different rice cultivars.

When evaluating the dry matter mass of the seedlings, there was a difference between the temperatures (25 and 17°C). At 25°C, the combination of products (dietholate+fipronil+carboxin+thiram) together with the control treatment differed from the other treatments, with the lowest values (Table 5).

Table 5: Accelerated ageing test, aerial part length, root length and dry matter mass for rice seeds treated and subjected to different temperatures.

Treatments	Accelerated ageing (%)		Air Part Length (cm)		Root length (cm)		Seedling dry matter mass (g)	
	25 C⁰	17°C	25 C⁰	17 Cº	25°C	17 C⁰	25 C⁰	17°C
T1-sem treatment	77 ab[1] A[2]	71 bA	10,05 b A[12]	0,65 bB	10,35 ns[4]	1,52 abB	0,0752 bc A[12]	0,0195 aB
T2-thiamethoxam	74 abcA	74 abA	9,76 bA	0,59 bcB	9,57	1,09 abcB	0,0820 abA	0,0147 aB
T3-dietholate	63 cA	68 bA	9,10 bA	0,35 bcB	10,35	0,54 bcB	0,0805 abA	0,0032 bB
T4-fipronil	83 aA	68 bB	9,05 bA	0,61 bcB	9,10	1,39 abB	0,0822 abA	0,0145 aB
T5-GA₃	83 aA	83 aA	13,88 aA	1,56 aB	10,07	1,95 aB	0,0940 aA	0,0127 aB
T6-carboxin + tirani	67 bcA	43 cB	9,32 bA	0,60 bcB	10,32	1=07 abcB	0,0790 bA	0,0142 aB
T7-DFCT[3]	64 cA	29 dB	6,60 cA	0,14 cB	9,05	0,14 cB	0,0615 cA	0,0012 bB
CV (%)		7,9		15,05		20,6		15,03

[1] Averages with different capital letters in the row differ by student's t-test (p≤0.05).

[2] Averages with different lower-case letters in the column differ by the Tukey test (p≤0.05)

[3] Dietolate, Carboxin + thiram - fipronil combinations.

[4] Not significant

The differences between the treatments and the control became less evident for the test at 25°C. Compared to the seeds subjected to a temperature of 17°C, the seeds that received the treatments with dietholate and the combination of dietholate with fungicide and insecticide also represented a reduction in the dry matter mass of the seedlings.

Experiment 2 - Effect of seed treatment on the initial establishment of irrigated rice at different sowing times.

For phytotoxicity, there was an interaction between the herbicide treatments applied and between the sowing times (Table 6), with no effect of seed treatment. For the phytotoxicity evaluations before the recommended time (1st time), at 14 and 21 DAH, there was a difference between the herbicides. The data from the phytotoxicity assessment at 28 DAH was not presented because the plants no longer showed signs of toxicity from the herbicide.

This result was also observed by Martini (2014) who, at seven and 28 DAH, also observed no symptoms of injury in rice plants when sown before the recommended time; the author used some herbicides, including bispiribaque sodium.

14

Some factors can affect the relationship between phytotoxicity and temperature, such as the plant species evaluated, the rate of detoxification and the herbicide used (Geier, Stahlman, Hargett, 1999; Koeppe et al., 2000; Mccullough and Hart, 2006). The herbicide bispiribaque-sodium showed the greatest phytotoxicity compared to the control treatment and profoxydim in the first sowing season (Table 6).

Table 6. Phytotoxicity in irrigated rice crops at two sowing times, as a function of herbicide treatments at 7, 14 and 21 days after herbicide application FAEM/UFPel - Capao do Leao, RS, 2013/14.

Herbicide	-	Phytotoxicity (%) 07 DAH[3]	
	Dose		
	(g i.a. ha)$^{-1}$	1^{a} Season	2nd Season
T1-control	-	0.0 b A[21]	0.0 cA
T2-bispiribach-sodium[4]	50	9 16 aB	27 08 bA
T3-profoxidim[4]	170	10 83 aB	53 90 aA
CV(%)		48,4	36,90
		Phytotoxicity (%)	
Herbicide	Dose	14 DAH[3]	
	(g i.a. ha)$^{-1}$	1 Season^{a}	2nd Season
T1-control	-	0.0 c A[21]	0.0 cA
T2-bispiribach-sodium4	50	17.7 aA	30.2 bA
T3-profoxidim[4]	170	8.9 bB	39.7 aA
CV(%)		45,9	41,2
		Phytotoxicity (%)	
Herbicide	Dose	21DAH[3]	
	(g i.a. ha)$^{-1}$	la Època	2nd Season
T1-control	-	0.0 c A[21]	0.0 bA
T2-bispiribach-sodium[4]	50	15.5 aA	15.0 aA
T3-profoxidim	170	11.8 bB	17.0 aA
CV(%)		45,4	43,5

1Means with different capital letters in the row differ by the Student's t-test (p≤0.05). 2Means with different lower-case letters in the column differ by Tukey's test (p≤0.05).

[3] Days after herbicide application.

[4] Post-emergence applications (v3-v4).

In the evaluation of phytotoxicity of rice plants, when sown at the recommended time (2nd Season), there was greater phytotoxicity of the ACCase inhibitor herbicide at seven and 14 DAH compared to the control. At 21 DAH, no difference was detected between the herbicides in the same season (October). When evaluations are carried out in the field, adverse conditions such as temperature, solar radiation and rainfall are fundamental to the results.

Air temperature is one of the main environmental variables to be observed when applying herbicides, especially when applying herbicides from the ACCase family, which are related to crop and weed control.

This effect was observed by Medd et al. (2001), who evaluated the effects of the maximum air temperature on the day of application and the sum of the minimum daily air temperatures of the seven days prior to the application of clodinafop-propargyl. It was found that low air temperature conditions at the time of or after application of the herbicide delayed responses in Avena spp. In the year in which the experiment was conducted (2013/14), the average temperature in the month of application of the treatments was contrasting. It was observed that after the application of the herbicides, temperature and radiation rose and then fell when the crop was grown before the recommended time (1st season). On the other hand, when the treatments were applied at the recommended time (2nd Season), the average air temperature fell and then rose, while solar radiation rose after the herbicides were applied (Figure 3).

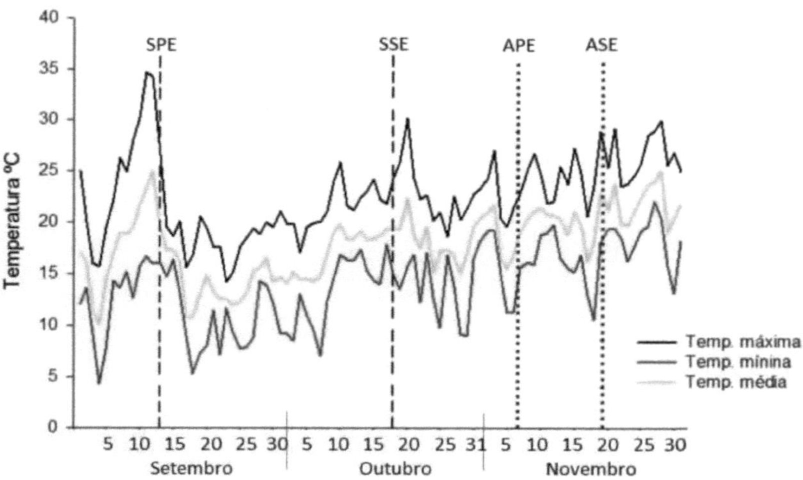

Figure 3: Maximum, minimum and average temperatures in September, October and November during sowing and herbicide application in 2013/14. SPE: Sowing first season; SSE: Sowing second

season; APE: Application first season; ASE: Application second season. FAEM/UFPel, Capao do Leao, RS, 2013.

For herbicides that inhibit the ACCase enzyme, their effectiveness is determined by an increase in air temperature, up to a certain limit. Low-temperature stress increases the wax content of leaves and decreases plant metabolism, resulting in lower absorption and translocation of the product (Cieslik, 2013).

The selectivity of rice to herbicides that inhibit the enzyme ACCases generally occurs at the level of the site of action, i.e. through the insensitivity of ACCase (Oliveira Jr. 2011). According to Dan Hess, (1994) there is no difference in absorption, translocation or metabolism between plant species.

At high temperatures, the effect of ACCase inhibiting herbicides can be reduced due to greater detoxification (Cieslik, 2013). The detoxification of herbicides by plants has an impact on selectivity for crops and weeds because it inactivates the herbicide molecule (Vidal, 2002; Cobb and Reade, 2010). Tolerance conferred by herbicide detoxification is generally associated with the cytochrome P450 monooxygenase and glutathione-S-transferase (GST) enzyme families (Reade, Milner and Cobb, 2004).

With regard to the effect of phytotoxicity on wheat plants, the application of ALS-inhibiting herbicides has led to an increase in injury symptoms as a result of higher temperatures (Geier, Stahlman, Hargett, 1999) and (Hoskins et al., 2005), corroborating the results of this study.

The possible explanation for the phytotoxicity at seven and 14 DAH in the second season is due to the temperature changes between the day of application and the days after application. The weather conditions during the herbicide application period at the recommended time, when there were days of low temperatures (Figure 6), favored an increase in symptoms and the death of rice plants.

The possible physiological changes caused by the temperature may have increased the susceptibility of the leaf tissue to the action of the herbicide, increasing the absorption of the product on the day of application. As the temperature rose, the symptoms of phytotoxicity became more visible, causing chlorosis, necrosis and the death of some plants (Figure 4).

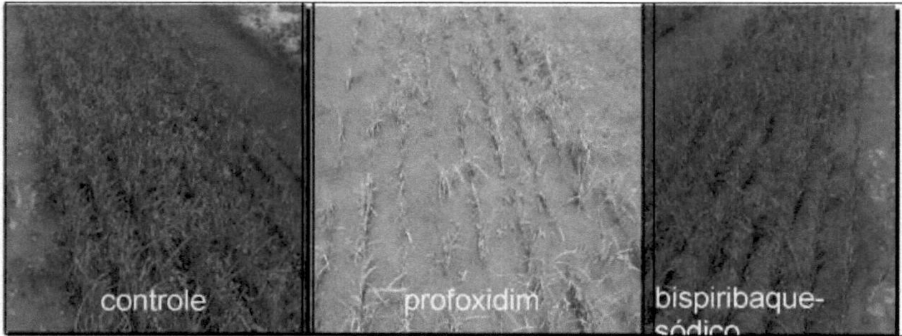

Figure 4. 14 DAH phytotoxicity assessment of rice plants sown at the recommended time (October) in relation to herbicides. FAEM/UFPel, Capao do Leao, RS, 2013/14.

Light intensity affects the activity of ALS-inhibiting herbicides in plants (Xie, Hsiao, Quick 1996; Camargo et al., 2012; Maciel et al., 2011), but the positive or negative effect is not clear from the literature.

Although it is considered a controversial factor, solar radiation must be taken into account in relation to the absorption of ACCase inhibiting herbicides. Normally, high levels of irradiance favor the activity of these herbicides (Kells et al., 1984; HattermanValenti et al., 2006).

This factor can be seen in the days after the application of the herbicides, where there was a significant decrease and subsequent increase in the rate of solar radiation, as can be seen in Figure 5. The rate of solar radiation two days after the application of the herbicide reached approximately 200 μmol m^{-2} s^{-1} , this condition may have favored the appearance of symptoms in the rice plants, due to the possible reduction in detoxification and unfavorable conditions for their metabolism.

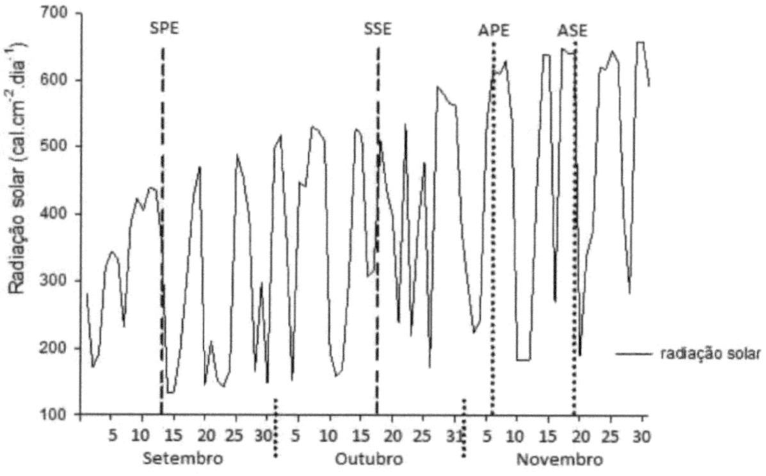

Figure 5: Solar radiation in September, October and November during sowing and herbicide application in 2013/14. SPE: First season sowing; SSE: Second season sowing; APE: First season application; ASE: Second season application. FAEM/UFPel, Capao do Leao, RS, 2013.

After applying the herbicide, there was a decrease in the rate of solar radiation, from approximately 600 cal. cm^{-2} day^{-1} to 200 cal. cm^{-2} day^{-1} , which may have resulted in greater absorption of the herbicide. This fact was also observed by (Xie, Hsiao and Quick 1996), in which lower radiation rates resulted in greater absorption of the herbicide[14] C-imazametabenz-methyl in Avena fatua.

On the other hand, work with the herbicide[14] C-imazetapir on herbicide-tolerant red rice plants (Oryza sp.) kept in growth chamber conditions showed superior absorption of the herbicide under higher light intensities than under conditions of reduced irradiance (Camargo et al., 2012).

Low rates of solar radiation mainly compromise photosynthesis, which directly influences the translocation of the product in the plant. This process was observed by (Hill & Stobbe, 1978) in Avena fatua, where the decrease in irradiance reduced the detoxification of the herbicide via conjugation, this effect favoring the control efficiency of the herbicide on the plant.

The influence of solar radiation on the application of herbicides that inhibit the enzyme ACCase, was highlighted in work with the poacea Setaria fabari where plants kept under high light intensity (812 $\mu mol\ m^{-2}\ s^{-1}$) showed fluazifop-P absorption approximately 25% higher compared to those under radiation conditions of (390 $\mu mol\ m^{-2}\ s^{-1}$) and 50% higher when compared to low light intensity (145 $\mu mol\ m^{-2}\ s^{-1}$) (Hatterman-Valenti et al., 2006).

Light affects some physiological processes in plants, which influence the absorption and translocation of ACCase inhibiting herbicides (Hull et al., 1982). Among other physiological and morphological factors, high irradiance favors the synthesis of photoassimilates, which is necessary for the transport of the herbicide in the plant (Wanamarta & Penner, 1989).

When evaluating the initial plant stand, there was a difference between the sowing times, but there was no difference between the herbicides before the recommended time (Season 1). At the recommended time, there was a difference between the herbicides, with profoxydim being the herbicide that most reduced the initial plant stand (Table 7).

After the rice was sown in September, the minimum air temperature in the month averaged 11°C and the average maximum temperature did not exceed 16°C. This condition of temperature stress lasted into the following month, with the average minimum and maximum temperatures being 13°C and 17.5°C respectively in October (Figure 3).

This condition was detrimental to the good initial development of the crop, which may explain the difference between the evaluation of the initial stand between the two seasons (Table 7). When the crop was sown at the recommended time (October), the period of exposure to low temperatures was shorter, as the November averages were higher than the previous months (Figure 3).

In the evaluation of the number of tillers, there was a difference between the sowing times and between the herbicides for the 1[a] sowing time, the herbicide profoxidim differed from the control treatment, and did not differ from bispiribaque sodium (Table 7).

Table 7. Initial stand and number of tillers in the irrigated rice crop at different sowing times, as a function of herbicide treatments. FAEM/UFPel - Capao do Leao, RS, 2013/14.

| Herbicide | Initial stand (plants linear meter) | | |
| | Dose | | |
	$(g\ i.a.\ ha\ )^{-1}$	1[a] Season	2nd Season
T1-control	-	27.38 a[2] B[1]	76.8 aA
T2-bispiribach-sodium	50	25.8 aB	72.0 aA
T3-profoxidim	170	25.4 aB	49.4 bA
CV(%) 47.8			

| Herbicide | Tillers (plants linear meter) | | |
| | Dose | | |
	$(g\ i.a.\ ha\ )^{-1}$	la Època	2nd Season
T1-control	-	172.4 a[2] B[1]	211.2 aA
T2-bispiribach-sodium	50	158.5 abB	231.6 aA
T3-profoxidim	170	145.7 bB	201.3 aA
CV(%) 24.9			

[1]Means with different capital letters in the line differ by the Student's t-test ($p \leq 0.05$).

[2]Means with different lower-case letters in the column differ according to the Tukey test ($p \leq 0.05$).

It is worth noting the recovery of the plants that received the herbicide profoxidim at the recommended time. This result can probably be attributed to the detoxification mechanism being able to detoxify itself from the effects of the herbicide (Martini, 2014). Another point is the ability of rice plants to tiller, Yoshida (1981) states that rice plants can compensate for the smaller stand by emitting a greater number of stalks.

The period between sowing and the emergence of the rice plants before the recommended time, in addition to the occurrence of low temperatures, was greatly affected by the conditions of excessive humidity, a factor that may have contributed to the reduction in the initial stand of rice plants (Figure 6).

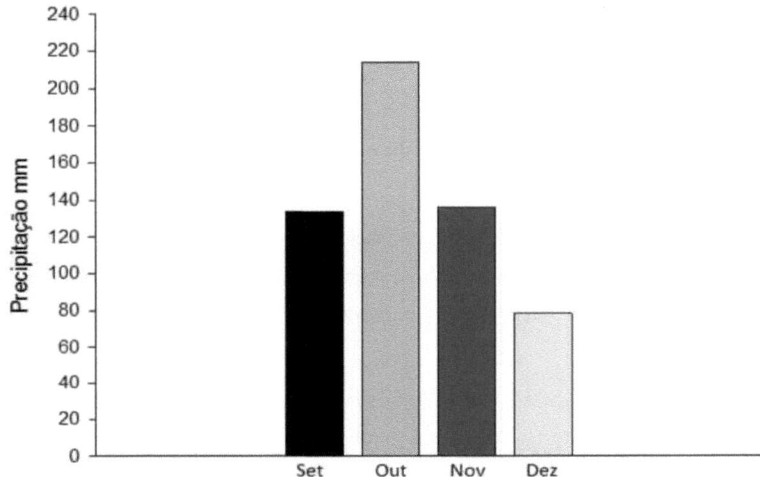

Figure 6. Rainfall in September, October, November and December during sowing and herbicide application in 2013/14. FAEM/UFPel, Capao do Leao, RS, 2013.

For the rainfall condition, there was an increase in rainfall in October, when the first season's rice plants were in full development. The rainfall in October was higher than the normal rainfall for the month. This may also have contributed to the poor development of the rice plants.

In the evaluation of panicles/m^{-2} , grains/panicle and weight of 100 grains, there was only a difference between the sowing times (Table 8).

Table 8. Panicles (m^{-2}), grains/panicle and weight of 100 grains in the irrigated rice crop at different sowing times. FAEM/UFPel - Capao do Leao, RS, 2013/14.

Sowing time	Panicula/ m^2	Grains/panicula	Weight 100 grains (g)
1st Season (September)	93,16 b[1]	100,18 a	2,21 b
2nd Season (October)	102 26 a	92 21 b	2 39 a
CV(%)	24 9	22 3	154

[1]Means with different lower-case letters in the column differ by the Student's t-test (p≤0.05).

This result is directly related to the grain yield at harvest. In general, the lower number of tillers in the first season compared to the second season led to a lower number of panicles, which may explain the higher number of grains/panicle in which the plant may have compensated for this difference.

The evaluation showed a direct relationship between plant stand, number of tillers and number of panicles m2, with the second season having the highest number of panicles and also the highest plant population.

In an experiment conducted by Martins Filho (2012), a study with nitrogen applications highlighted the increase in tillering in rice plants, which increased the number of panicles and, consequently, the final grain yield.

The components of a crop's grain yield include three main factors: the number of panicles; the number of grains/panicle and the weight of grains. In the second season, there were more tillers, resulting in a higher number of panicles, but fewer grains/panicle. Even with this result, the second season showed higher productivity.

The observed results of lower grains/panicle in the second season, the higher grain weights, allow us to infer that these results explain the fact that the second season represents higher productivity compared to the first season.

As for average yields, the results obtained in the experiment show higher grain yields in the recommended season (October) of 11,000 kg ha^{-1} , and yields before the recommended season were around 9,000 kg ha^{-1} , in this evaluation the herbicide factor had no statistical difference (Figure 7).

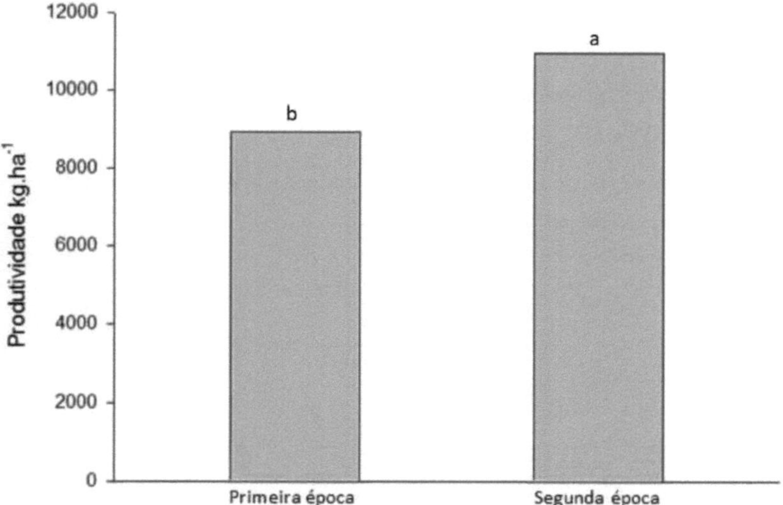

Figure 7. Productivity of irrigated rice in two sowing seasons. FAEM/UFPel, Capao do Leao, RS, 2013/14.

Sowing time is one of the main factors that define rice grain productivity. Choosing the ideal time is an important decision and depends on various factors such as environmental conditions, soil type, cultivar and the incidence of weeds in the area. In the cultivation of irrigated rice, the growing season is limited to the period in which the factors temperature and solar radiation are available in sufficient quantities to allow the plant to develop fully.

Based on the above, it can be inferred that seed treatment with gibberellic acid positively influenced rice seeds at the optimum temperature (25°C) and suboptimum temperature (17°C). When the seeds were treated with GA3 at low temperatures, it had a positive influence on seedling vigor.

Treatment with dietholate and the combination of dietholate+fipronil+carboxin+tiram negatively influenced germination and vigor tests at both temperatures, reducing the physiological performance of rice seeds in the laboratory. When taken to the field, the seed treatments did not influence the factors analyzed, being dependent on the application of the herbicides and the sowing time.

Given the weather conditions, temperature was the factor that most influenced phytotoxicity in the crop, with the ALS-inhibiting herbicide showing greater toxicity when sown at the beginning of September. When sown at the recommended time, the plants showed greater symptoms of injury when they received applications of the ACCase inhibiting herbicide.

Even though the rice plants showed phytotoxicity, they were able to detoxify the herbicide. This can be analyzed because there were more tillers, resulting in a greater number of panicles, giving higher yields.

BIBLIOGRAPHICAL REFERENCES

ALMEIDA, A.S. et al. Bioactivator in the physiological performance of carrot seeds. Revista Brasileira de Sementes, v.31, p.87-95, 2009.

ALMEIDA, A, S. et al. Bioactivator in the physiological performance of rice seeds. Revista brasileira de sementes, v.33, p. 501-510, 2011.

ALMEIDA, A, S. et al. Thiamethoxam: An Insecticide that Improves Seed Rice Germination at Low Temperature. Insecticides - Development of Safer and More Effective Technologies. Intech, v.14, p.417-425, 2013.

AOAC - ASSOCIATION OF OFFICIAL AGRICULTURAL CHEMISTS, Official methods of analysis. 10 ed. Washington: Editorial Board, 1965. 909 p.

BEVILAQUA, G.A.P. et al., Performance of rice seeds treated with growth regulators. I. Effect on field emergence. Revista Brasileira de Sementes. v. 15, p. 75-80. 1993.

BRAZIL. Ministry of Agriculture, Livestock and Supply. Rules for seed analysis. Brasilia, 2009. 399p.

CAMARGO, E.R. et al., Interaction between saflufenacil and imazethapyr in red rice (Oryza ssp.) and hemp sesbania (Sesbania exaltata) as affected by light intensity. Pest Management Science, v.68, p.1010-1018, 2012.

CASTRO, P.R.C. et al. Action of plant regulators on the development, nutritional and anatomical aspects and productivity of bean (Phaseolus vulgaris) cv. Carioca. Anais... da Esalq, v.47, p11-28, 1990.

CIESLIK, L.F. et al., Environmental factors affecting the efficacy of accase inhibiting herbicides: a review. Planta Daninha, Viçosa-MG, v. 31, p. 483-489, 2013.

COBB, A.H.; READE, J.P.H. Herbicides and plant physiology. 2nded. Newport-UK: Harper Adams University, Wiley-Blackwell, 2010. 286 p.

CONAB. Rice - Brazil. Monitoring the Brazilian harvest. Twelfth survey September 2014. Available at: <www.conab.gov.br> Accessed on: October 2014.

CRUZ, R.P,; DUARTE, I.T.L.; CABREIRA, C. Inheritance of rice cold tolerance at the seedling stage. Science Agricola, v.67, p.669-674, 2010.

CUNHA, R.; CASAL, V.W. Effect of growth regulating substances on lettuce seed germination. Revista Brasileira de Fisiologia Vegetal, v.1, p.121-132, 1989.

DAN HESS, F., Mode of action of lipid biosynthesis inhibitors (graminicides ACCase inhibitors). In: Purdue University, (Ed,). Herbicide Action Course. West Lafayette, USA: CRC Press, p.201-216, 1994.

DERIDDER, B.P.; DIXON, D.P.; BEUSSMAN, D.J. et al. Induction of glutathione S- transferases in arabidopsis by herbicide safeners. Plant Physiology, v.130, p.14971505, 2002.

DIAS, A. D., GOMES, A.S., Effect of seed treatment with gibberellic acid on the performance of irrigated rice. Revista Brasileira de Agrociência, v.1, p. 97-102, 1995.

EMBRAPA. Sistema Brasileiro de Classificaçâo de Solo. 2ª ed., Rio de Janeiro: EMBRAPA, 306 p. 2012.

FLORES, I.F., Treatment of seeds with gibberellic acid and growth of rice seedlings (Oryza sativa L.). Revista da FZVA, Uruguaiana, v. 9, p. 73-78. 2002.

FORD, K.A. et al. Neonicotinoid insecticides induce salicylate-associated plant defense responses. Proceedings of the National Academy of Sciences, v.107, p.17527-17532, 2010.

FREITAS, T.F.S. et al. Irrigated rice productivity and nitrogen fertilization efficiency influenced by sowing time. Revista Brasileira de Ciência do Solo, v.32, p.2397-2405, 2008.

GEIER, P.W.; STAHLMAN, P.W.; HARGETT, J.G. Environmental and application effects on MON 37500 efficacy and phytotoxicity. Weed Science, v.47, p.736-739, 1999.

HASHIMOTO, M; KOMATSU, S. Proteomic analysis of rice seedlings during cold stress. Proteomics, v.7, p.1293-1302, 2007.

HATTERMAN-VALENTI, H. M.; PITTY, A.; OWEN, M.D. K. Effect of environment on giant foxtail (Setaria faberi) leaf wax and fluazifop-P absorption. Weed Science, v. 54, p. 607-614, 2006.

HILL, B. D.; STOBBE, E. H. Effect of light and nutrient levels on 14C-benzoylprop ethyl metabolism and growth inhibition in wild oat (Avena fatua L.). Weed Research, v. 18, p. 223-229, 1978.

HOSKINS, A.J.; YOUNG, B.G.; KRAUSZ, R.F.; RUSSIN, J.S. Control of italian ryegrass (Lolium multiflorum) in winter wheat. Weed Technology, v.19, p.261-265, 2005.

HIRASE , K.; MOLIN, W.T. Sulfur assimilation in plant weed control: Potential targets for novel herbicides and action sites of certain safeners. Weed Biology and Management, Stoneville, v.3, p.147-157, 2003.

HULL, H. M.; DAVIS, D. G.; STOLZENBERG, G. E. Action of the adjuvants on plant surfaces. In: HODGSEN, R. H. Adjuvants for herbicides. Champaign: Weed Science Society of America, 1982. p. 26-67.

IRGA-INSTITUTO RIOGRANDENSE DO ARROZ, 2012. Available at <http://www.irga.rs.gov.br/uploads/anexos/1337000441.Semeadura_e_Colheita_do_Arr oz_no_RS.pdf.> Accessed on: May 2014.

ISTA. International Seed Testing Association, 23 Mar. 2005. Specials. Accessed on September 30, 2013. Online. Available at: <http://seedtest.org/en/seed-testing-international- _content---1-- 1085.html> Accessed on: May 2014.

JURSiK, M.; SOUKUP, J.; HOLEC, J. Herbicide mode of actions and symptoms of plant injury by herbicides: inhibitors of acetolactate synthase (ALS inhibitors). Listy Cukrovarnické a Reparské, v.126, p.376-379, 2010.

KAPOOR, N. et al. Physiological and Biochemical Changes During Seed Deterioration in Aged Seeds of Rice (Oryza sativa L.). American Journal of Plant Physiology, v.6, p.28-35, 2011.

KING, R.W., et al, Gibberellins in relation to growth and flowering in Pharbitis nil Chois. Canberra, Australia. Plant Physiology. v.84, p.1126-1131, 1987.

KOEPPE, M.K. et al., Basis of selectivity of the herbicide rimsulfuron in maize. Pesticide Biochemistry and Physiology, v. 66, p.170-181, 2000.

KROHN, G. N.; MALAVASI, M. M. Physiological quality of soybean seeds treated with fungicides during and after storage. Revista Brasileira de Sementes, Londrina, v. 26, p. 91-97, 2004.

KELLS, J. J.; MEGGITT, W. F.; PENNER, D. Absorption, translocation, and activity of fluazifop-butyl as influenced by plant growth stage and environment. Weed Science, v. 32, p. 143-149, 1984.

LEVITT, J. Introduction to plant physiology. 2.ed. Saint Louis: The C.V. Mosby Company, 1974. 447p.

LIMA, M.G.S. et al. Physiological quality of rice seeds subjected to salt stress. Revista Brasileira de Sementes, v.27, p.54-61, 2005.

LILGE, C.G. et al. Performance of rice seeds of different cultivars in the presence of the herbicide glufosinate ammonium. Revista Brasileira de Sementes. Pelotas, v.25, p.82-88, 2003.

MACEDO, W.R; CASTRO, P.R.C. Thiamethoxam: Molecule moderator of growth, metabolism and production of spring wheat. Pesticide Biochemistry and Physiology, v.100, p. 299-304, 2011.

MACIEL, C.D.G.; et al., Development of lawns submitted to the application of growth retardants in different light conditions. Planta Daninha, v.29, p.383-395, 2011.

MCCULLOUGH, P.E.; HART, S.E. Temperature influences creeping bentgrass (Agrostis stolonifera) and annual bluegrass (Poa annua) response to bispyribac-sodium. Weed Technology, v.20, p.728-732, 2006.

MARCOS-FILHO, J. et al., Effects of seed size on germination, vigor and yield of maize (Zea mays L.). Anais... da ESALQ, Piracicaba, v.34, p.327-337, 1977.

MARCOS FILHO, J. Fisiologia de sementes de plantas cultivadas. Piracicaba: Fealq, 495p, 2005.

MARTINI, L.F.D., Selectivity of herbicides on irrigated rice under cold stress conditions. 2014. 118f. Thesis (Doctorate in Plant Health) - Faculty of Agronomy. Federal University of Pelotas. Pelotas/RS.

MARTINS, J.M.C. et al., Relationship between grain yield and panicle production in hybrid rice cultivars subjected to base and top dressing doses of n. Revista Tècnico Cientifica (IFSC), v. 3, p. 744, 2012.

MEDD, R. W. et al. Determination of environment specific dose-response relationships for clodinafop-propargylon Avena spp. Weed Resesearch, v. 41, p. 351-368,2001.

MERTZ, L.M. et al., Physiological changes in rice seeds exposed to cold during germination. Revista

Brasileira de Sementes, v. 31, p.254-262, 2009.

MISTURA, C.C. et al., Influence of the seed protectant diethylphenylphosphorothioate on rice seedlings (Oryza sativa L.) Revista Brasileira de Agrociência. v.14, p.231238, 2008.

PESKE, S.T., BEVILAQUA, G.A.P., Treatment of rice seeds with gibberellic acid. In: Reuniao da Cultura do Arroz Irrigado, 19, Camboriù, SC. 1991. Anais... Camboriù, 1991.

PIGATO, F.J. et al., Effect of gibberellic acid on the productivity of upland rice. UNICiência v.14, p.225-240, 2010.

POPINIGIS, F. Seed Physiology. Brasilia: ABEAS, 1985. 289p.

READE, J.P.H.; MILNER, L.J.; COBB, A.H. A role for glutathione S-transferases in resistance to herbicides in grasses. Weed Science, v.52, p.468-474, 2004.

ROGIS, C. et al., Enhancing germination of eastern gama grass seed with stratification and giberelic acid. Crop Science v. 44, p. 549-552, 2004.

SALGADO, F. H. M. Maize seed germination treated with insecticides. Journal of Biotechnology and Biodiversity, v. 4, p. 49-53, 2013.

SCHWECHHEIMER, C. Understanding gibberellic acid signaling-are we there yet. Plant Biology, v. 1, p. 9-15, 2008.

SILVA, W.R. & MARCOS-FILHO, J. Effects of maize seed weight and size on germination and vigor in the laboratory. Revista Brasileira de Sementes, v.1, p.39-52, 1979.

BRAZILIAN SOCIETY OF WEED SCIENCE - SBCPD.

Procedures for setting up, evaluating and analyzing experiments with herbicides. Londrina: p. 45, 1995.

SOSBAI - SOUTH-BRAZILIAN SOCIETY OF IRRIGATED RICE. Irrigated Rice: SOSBAI. Technical Research Recommendations for Southern Brazil, 28. Technical Meeting on Irrigated Rice Culture. Porto Alegre: SOSBAI, p. 188, 2012.

SOUZA, P.R. de, MENEZES, V.G., Gibberellic acid in the treatment of irrigated rice seeds. Lavoura Arrozeira, Porto Alegre, v.44, p.3-4, 1991.

STEINMETZ, S., BRAGA, H. J. Zoning of irrigated rice by sowing times in the states of Rio Grande do Sul and Santa Catarina. Revista Brasileira de Agrometeorologia, v.9, p.429-438, 2001.

SUGE, H. Stimulation of oat and rice mesocothyl growth by ethylene. Plant and Ceil Physiology, v.12, p.831-837, 1971.

VIDAL, R. A.; WINKLER, L. M. Weed resistance: selection or mutation induction by acetolactate synthase (ALS) inhibitor herbicides. **Pesticidas: Revista de Ecotoxicologia e Meio Ambiente**, v. 12, p. 31-42, 2002.

VILA-AIUB, M. et al. Glyphosate resistance in Sorghum halepense and Lolium rigidum is reduced at suboptimal growing temperatures. **Pest Management Science**, v.69, p.228-232, 2012.

XIE, H. S. et al. Influence of water stress on absorption, translocation, and phytotoxicity of fenoxaprop-ethyl and imazamethabenz-methyl in Avena fatua. **Weed Research**, v. 36, p. 65-71, 1996.

WANAMARTA, G. D.; PENNER, D. Foliar absorption of herbicides. **Weed Science**, v. 4, p. 215-231, 1989.

YAZBEK JR., W.; FOLONNI, L.L. Effect of seed protectant on herbicide selectivity in cotton crop (Gossypium hirsutum L.). **Revista Ecossistema**, v.29, p.33-38, 2004.

YOSHIDA, S. **Fundamentals of rice crop science**. Los Banos: IRRI, 277 p. 1981.

Biochemical changes in irrigated rice in response to stress caused by herbicides and temperature[6]

Biochemical changes in rice in response to stress caused by herbicides and low temperatures

ABSTRACT - The selectivity of herbicides in irrigated rice is influenced by several aspects, including the environmental conditions after application. The aim of this study was to evaluate and characterize the biochemical parameters of oxidative stress in irrigated rice subjected to different temperatures, seed treatments and the application of herbicides used in rice cultivation. The rice seeds were sown at two temperatures: 25°C and 17°C. The seed treatments used were: thiamethoxan, dietholate, fipronil, gibberellic acid, carboxin+thiram, dietholate+fipronil+carboxin+thiram and control (no application), and the herbicides applied were profoxidim, bispiribaque-sodium and a control (no herbicide application), The variables assessed were total chlorophyll and carotenoid levels, activity of the enzymes catalase, ascorbate peroxidase and superoxide dismutase, lipid peroxidation and hydrogen peroxide levels. Leaf samples were taken five days after the herbicides were applied. There was no difference between the variables analyzed - seed treatment and herbicides. The results were dependent on the herbicides and show that their application alters the metabolism of rice at a temperature of 25°C. The herbicides profoxidim and bispiribaque sodium cause oxidative stress, but the crop activates its antioxidant defense system through the action of the enzyme superoxide dismutase. When subjected to low temperatures (17°C), the application of the different hebicides caused less oxidative damage; the low metabolism of the plants due to the low temperature may explain this result. Therefore, the data show that exposing rice to different temperatures and herbicides induces oxidative stress, as well as changes in the activities of antioxidant enzymes.

Keywords: oxidative stress, low temperature, Oryza sativa, selectivity

[1]Master's student, Post-Graduate Program in Plant Health, Federal University of Pelotas - UFPel, CAPES scholarship holder;[2] Adjunct Professor, Ph.D., Department of Plant Health, UFPel, P.O. Box 354, 96010-900 Pelotas- RS, Brazil, <laavilabr@gmail.com>;[3] Post-Doctorate, Federal University of Pelotas, UFPel, Brazil. Scholarship from: Coordenaçao de Aperfeiçoamento de Pessoal de Nivel Superior, CAPES[4] Master's student, Postgraduate Program in Plant Physiology, Universidade Federal de Pelotas - UFPel, CAPES Fellow;[5] Adjunct Professor, Department of Plant Physiology, UFPel, Caixa Postal354, 96010-900 Pelotas- RS, Brazil, <sdeuner@yahoo.com.br>;[6] Article formatted according to the standards of the journal Planta Daninha.

INTRODUCTION

Because it is grown in different regions, rice is subject to various environmental conditions, but when compared to other cereals, such as wheat, rice is more sensitive to low temperature stress (Okuno, 2003). In the state of Rio Grande do Sul, most of the rice cultivars used in the Iavouras are of tropical origin, and the occurrence of cold weather becomes an aggravating factor when it comes to the cereal's sensitivity to abiotic stress. This factor causes serious damage to the initial establishment of the crop, resulting in a reduction in the initial stand and consequently the growth and development of weeds (Cruz, 2001).

As a result of changes in temperature in recent years, the productivity of irrigated rice in the state has suffered instability, with the occurrence of low temperatures being one of the main factors for oscillations in the crop's productivity (Steinmetz et al., 1996). The temperature range for good rice development is between 25 and 30°C (Yoshida, 1981), and temperatures below this range can cause damage and stress to the crop.

The most serious damage to the rice crop caused by low temperatures is related to the low fecundity of the spikelets, which is due to their sterility because of the unviability of the pollen. During flowering, the cold damages the dehiscence of the anthers and also the growth of the pollen tube (Yoshida, 1981, Cruz and Milach, 2000).

The application of herbicides is also one of the factors that causes oxidative stress in rice plants. Herbicides applied to control weeds can also cause damage to the crop, resulting in phytotoxicity, reducing its growth and development (Nohatto, 2014).

The herbicides profoxidim and bispiribaque-sodium are selective herbicides used in rice cultivation, applied in post-emergence to control some weeds such as rice grass (Echinochloa sp.) in rice cultivation.) in rice cultivation, but when some selective herbicides are applied in adverse conditions such as early sowing, where temperatures are variable, many herbicides in low temperature conditions can cause biochemical and physiological disturbances in the metabolism of plants, reducing the selectivity of the herbicide, which negatively affects their development and consequently their production (Song et al., 2007).

This increase in phytotoxicity in the rice crop was observed by Cassol, 2012, where herbicides such as propanil, clomazone and cyhalofop-butyl when applied to irrigated rice sown in September showed increased levels of phytotoxicity compared to the crop sown in November.

When the plant is in adverse conditions, several harmful reactions occur in its development, such as a decrease in photosynthesis, together with changes in its metabolism (Law & Crafts-brandner, 2001). Associated with these reactions is the production of reactive oxygen species (ROS) which can accumulate in the tissues. This mechanism results in oxidative damage to proteins, DNA and lipids, such as superoxide (O_2), hydrogen peroxide (H_2O_2) and hydroxyl radicals (OH^-) (Grohs, 2012).

Formed through normal metabolic reactions as by-products and allocated to different cellular compartments, ROS can be induced through stimuli depending on the environmental conditions in which the plants find themselves (Mittler, 2002). Under stable physiological conditions, ROS are removed by plant defense components, and these antioxidant components are often allocated to particular compartments. The ratio between the production of these molecules and their elimination depends fundamentally on various environmental factors (Grohs, 2012).

ROS cause lipid peroxidation, disintegration of membranes, damage to proteins and nucleic acids (Gomez et al., 1999), common effects in plants subjected to low temperature stress (Morsy et al., 2007). In the chloroplast, ROS cause a decrease in the activity of proteins such as ATP synthase and vacuolar ATPase, explaining the decrease in photosynthetic rate and the consequent decrease in growth (Neilson et al. 2010).

Through evolution, higher plants have developed numerous mechanisms to respond to stresses, such as the ability to alter their development in response to unfavorable external factors such as attacks by pests, pathogens and abiotic factors (Soares & Machado, 2007). Plants have an active detoxification system, with the action of antioxidant enzymes such as superoxide dismutase, peroxidase, glutathione S-transferase, ascorbate peroxidase and catalase (Foyer & Nector, 2000), as well as low molecular weight non-enzymatic antioxidants such as ascorbate and reduced glutathione.

The antioxidant system protects the integrity of membranes against the effects of ROS, enabling certain species to be more tolerant of adverse environmental conditions (Noctor & Foyer, 1998). The tolerance of rice to drought or cold is directly related to the capacity of the enzymatic antioxidant system (Guo et al. 2006).

The rice crop is sensitive to cold stress, and its exposure to this condition can cause various types of damage, especially to germination and initial establishment, resulting in yield losses.

In recent years, seed treatment has been seen not only as a tool for protecting seeds against pathogen attack. Many products such as thiamethoxan have been highlighted as bioactivators capable of improving the physiological quality of rice seeds (Almeida, et al. 2011). Therefore, the need for studies into the use of products for seed treatment, which can result in protection or improvements in the crop, to excel when subjected to atypical factors, is of great importance and interest to producers.

Biochemical variables and their behavior resulting from plant metabolic stress are important factors that trigger processes such as modulation of plant gene expression. Knowing how plants protect themselves against different stresses is important in order to obtain, with the help of genetic improvement programs, more resistant and/or tolerant agricultural varieties, which can consequently increase plant production and quality. There are several substances involved in inducing plant defense responses against many stresses that deserve to be investigated.

In view of the above, the aim of the study was to evaluate and characterize the biochemical parameters of oxidative stress in irrigated rice subjected to different seed treatments, temperatures and the application of herbicides used in irrigated rice cultivation.

MATERIAL AND METHODS

The work was divided into two experiments carried out in a growth chamber at different temperatures of 25 and 17°C, installed at the Eliseu Maciel Agronomy Faculty belonging to the Federal University of Pelotas, in the 2014 agricultural year.

The experimental units were made up of plastic cups with a capacity of 700 mL, containing soil collected from the A horizon of a solodic Haplic Eutrophic Planossolo (EMBRAPA, 2006), dried in open air and sieved through a 2.0 mm mesh sieve, using 600 g of soil per cup.

The design used was randomized blocks with subdivided plots, arranged in a factorial scheme, with four replications, where Factor A consisted of two air temperatures: Optimum: 25°C and Suboptimum = 17°C; Factor B consisted of seven seed treatments (Table 1), with the products being applied directly to the seeds with a pressurized valve, 24 hours before sowing and placed in plastic bags with a capacity of five liters, using 1 kg of seeds per bag. The volume of spray used was 1.5 L 100 kg^{-1} of seed, and for the control treatment only water was used.

Table 1. Products used and registered for seed treatment in rice cultivation. FAEM/UFPel - Capão do Leão, RS, 2014.

Treatments	Factor B: Active Ingredient in TS	Dose g a.i. 100 kg^{-1}
1	no application	---
2	thiamethoxam	140,0
3	dietholate	600,0
4	fipronil	62,5
5	Gibberellic acid	2,0
6	carboxin+tyram	60,0 + 60,0
7	dietolate+fipronil+carboxin+tyram	600,0+62,5+60,0+60,0

Factor C consisted of the herbicides bispiribaque-sodium and profoxidim and the control (Table 2). The herbicide treatments were applied at the point of emergence of the irrigated rice (3 to 4 leaves), following the manufacturer's recommendations for use.

Table 2. Registered and recommended herbicides for weed control in irrigated rice cultivation in RS and SC. FAEM/UFPel - Capão do Leão, RS, 2014

Active Ingredient	Registration dose of P.c	Method of application

	(g i.a. ha-)	
control	-	-
bispiribaque sodium	50	post-emergency
profoxidim	170	Post-emergence

Plant material and growing conditions

The soil used in the experiment had no history of herbicide application in the last five years, and was collected from the A horizon of the varzea area of the Centro Agropecuàrio da Palma (CAP) belonging to FAEM-UFPel, classified as a solodic eutrophic Hydromorphic Planosol (Pelotas Mapping Unit), which had the following characteristics: pHwater (1:1) = 5.1; CTC pH7 = 5.4 cmolc dm-3; Organic Matter = 1.2 %; clay = 15 %; texture = 4; Ca = 1.8 cmolc dm^{-3} ; Mg = 1 cmolc dm^{-3} ; Exchangeable Al = 0.2 cmolc dm^{-3} ; Available P = 4.3 mg dm^{-3} ; Exchangeable K = 30 mg dm$^{-3.}$ The chemical analysis was carried out in the Soil Analysis Laboratory of the UFPel Soil Department. After being filled with soil, the cups were

Figura 1. View of the experimental units during sowing. FAEM/UFPel, Capao do Leao, RS, 2014.

To apply the herbicides, a CO2-pressurized precision knapsack sprayer was used, equipped with a boom fitted with four 110-02 series flat fan jet tips, spaced 50 cm apart, calibrated to apply a spray volume of 150 L ha^{-1} , and the herbicides were applied post-emergence.

As for obtaining the plant material, samples of the aerial part of the plants were collected five days after the application of the herbicide treatments (DAT). To assess the effect of the treatments on biochemical parameters, four plants were collected per experimental unit, totaling 24 plants per treatment. Three leaves were taken from each plant for biochemical analysis. These samples were frozen in liquid nitrogen and immediately stored at -80 °C for later determination of hydrogen peroxide content, lipid peroxidation, antioxidant enzyme activity, chlorophylls, carotenoids and phenolic compounds.

Determination of hydrogen peroxide (H_2O_2) content and lipid peroxidation

Cell damage in the tissues was determined using the hydrogen peroxide (H_2O_2) content, as described by Loreto & Velikova (2001) and thiobarbituric acid reactive species (TBARS), via the accumulation of malonic aldehyde (MDA), as described by Health and Packer (1968). To carry out these analyses, 0.2g of leaves were macerated in liquid nitrogen, homogenized in 2 mL of 0.1% (w/v) trichloroacetic acid (TCA) and centrifuged at 14000 rpm for 20 minutes. For H_2O_2 quantification, aliquots of 0.2 mL of the supernatant were added to 0.8 mL of 10 mM phosphate buffer (pH 7.0) and 1 mL of 1M potassium iodide. The solution was left to stand for 10 minutes at room temperature and the absorbance was read at 390 nm. The concentration of H2O2 was determined using a standard curve and expressed in mM/g.

To determine TBARS, 0.5 mL aliquots of the supernatant, as described above, were added to 1.5 mL of 0.5% (w/v) thiobarbituric acid (TBA) and 10% (w/v) trichloroacetic acid and incubated at 90°C for 20 minutes. The reaction was stopped in an ice bath for 10 minutes. The absorbance was read at 535 nm, discounting the unspecific absorbance at 600 nm. The concentration of MDA was calculated using the absorbance coefficient of 155 mM cm^{-1} , and the results were expressed as nM MDA g^{-1} of MF.

Leaf photosynthetic pigments

The chlorophyll (a, b and total) and carotenoid contents were determined from a 0.1g sample of the aerial part macerated in a crucible in the presence of 5 mL of 80% (v/v) acetone. The material was centrifuged at 12,000 rpm for 10 minutes and the supernatant transferred to a 25 mL volumetric flask, topping up with 80% (v/v) acetone. The levels of total carotenoids, chlorophylls a, b and total were calculated using the formula of Lichtenthaler (1987) from the absorbance of the solution obtained by spectrophotometry at 647, 663 and 470 nm, and the results were expressed in mg g^{-1} of MF.

Preparation of the enzyme extract

To determine the enzymatic activity, 0.2 g of the sample was macerated using liquid nitrogen and 0.02 g of polyvinylpyrrolidone (PVPP). Then 900 μL of 200 mM phosphate buffer (pH 7.8), 18 μL of 10 mM EDTA, 180 μL of 200 mM ascorbic acid and 702 μL of ultrapure water were added and centrifuged at 14000 rpm at 4°C for 20 minutes. The supernatant was collected and used for further analysis. From this extract, the protein of the samples was quantified using the Bradford method (1976), the standard curve with globulin was drawn up and the results expressed in mg g^{-1} MF.

Determination of antioxidant enzyme activities

The activity of catalase (CAT; EC 1.11.1.6) was determined through the consumption of H2O2 (extinction coefficient 39.4 mM cm^{-1}). To do this, 1 mL of 200 mM potassium phosphate buffer (pH

7.0), 850 μL of ultrapure *water,* 100 μL of 250 mM hydrogen peroxide and 50 μL of extract were reacted. Absorbance readings at a wavelength of 240 nm were taken on a spectrophotometer (Ultrospec 6300 Pro UV/Visible - Amersham Bioscience) for 90 seconds at 7-second intervals.

The activity of ascorbate peroxidase (APX; EC 1.11.1.11) was determined according to Azevedo et al. (1998), with modifications, through the consumption of H2O2 (extinction coefficient 2.9 mM cm^{-1}). 1 mL of 200 mM potassium phosphate buffer (pH 7.0), 750 μL of ultrapure water, 100 μL of 10 mM ascorbic acid, 100 μL of 2 mM hydrogen peroxide and 50 μL of extract were used. Absorbance readings at a wavelength of 290 nm were taken on a spectrophotometer (Ultrospec 6300 Pro UV/Visible - Amersham Bioscience) for 90 seconds at 7-second intervals. For both CAT and APX activity, for calculation purposes, a decrease of one absorbance unit was considered to be equivalent to one active unit (AU). The activities of the total extract were determined from the amount of extract that reduced the absorbance reading by one AU, and expressed as AU mg^{-1} protein minute $^{-1}$.

The activity of superoxide dismutase (SOD; EC 1.15.1.1) was determined according to a methodology adapted from Peixoto (1999), by calculating the amount of extract that inhibited 50% of the NBT reaction and expressed as UA mg^{-1} protein minute^{-1}. In a test tube, 1 mL of 100 mM potassium phosphate buffer (pH 7.8), 400μL of 70 mM methionine, 20 μL of 10 μM EDTA, 390 μL of ultrapure water, 150 μL of 1 mM NBT, 20 μL of 0. 2 mM riboflavin and 20 μL of 0. 2 mM riboflavin were added,2 mM and 20 μL of extract incubated for 10 minutes in a 15-watt fluorescent lamp and the absorbance was read on a spectrophotometer (Ultrospec 6300 Pro UV/Visible - Amersham Bioscience) at a wavelength of 560 nm. For the purposes of calculation, the reaction blank was considered to be tubes containing no extracts, exposed and not exposed to light. Activity was determined by calculating the amount of extract that inhibited 50% of the NBT reaction and expressed as UA mg^{-1} protein minute^{-1} (Figure 2).

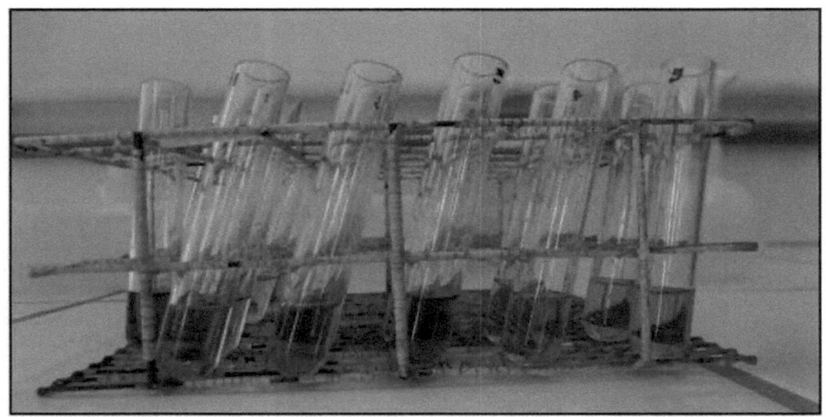

Figura 2. Superoxide dismutase (SOD) activity, FAEM/UFPel, Capâo do Leâo, RS, 2014.

The data obtained was analyzed for normality using the Shapiro-Wilk test and for homoscedasticity using the Hartley test, and then submitted to variance analysis ($p \leq 0.05$). After the analysis of variance, the t-test was used to compare the means in the event of a significant difference between the seed treatments Duncan's test ($P \leq 0.05$) was used to compare the herbicide treatments.

RESULTS AND DISCUSSION

Experiment 1- Biochemical changes in irrigated rice in response to the application of seed treatment and herbicides at a temperature of 25°C.

There was no interaction between seed treatment and herbicides. For the total chlorophyll concentration, profoxydim caused a reduction in the content of this pigment, which did not differ from the herbicide bispiribaque-sodium. The plants subjected to the control treatment (no herbicide application) had a higher chlorophyll content, which differed from the profoxidim treatment. This result indicates that the application of profoxidim has the ability to reduce the total chlorophyll content of rice plants at 25 °C (Table 3).

Table 3. Concentration of total chlorophyll (mg g^{-1} MF) and carotenoids (mg g^{-1} MF) five days after herbicide application on IRGA 424 rice leaves, subjected to a temperature of 25°C. FAEM/UFPel, Capao do Leao, RS, 2014.

Herbicides	Dose (g i.a. $^{ha-1}$)	Total Chlorophyll	Carotenoids
witness	-	3,55 a	0,39 a[1]
profoxidim	170	2,52 b	0,28 b
bispiribaque sodium	50	2.99 ab	0.33 ab
CV (%)		15,77	20,6

[1]Lowercase letters compare herbicides using the Tukey test ($p < 0.05$).

Chlorophyll molecules are the main pigments responsible for capturing light for photochemical reactions, present in the reaction centers of photosystems (Taiz and Zeiger, 2009). The decline of these compounds can compromise photosynthetic activity, resulting in damage to plant development (Ramesh et al., 2002).

There was an effect of the herbicide profoxydim reducing the chlorophyll content compared to the control, but it did not differ from the herbicide bispiribaque-sodium. The lower capacity of herbicides that inhibit the ACCase enzyme, such as profoxydim, to influence the behavior of this variable was also observed by Park et al. 1994, corroborating our results.

With regard to carotenoid content, the results were similar to those presented for total chlorophyll content. The herbicide bispiribaque-sodium did not differ from the herbicide profoxidim and the control (Table 3).

Carotenoids are accessory pigments in the absorption and transfer of radiant energy, and protect chlorophyll from photooxidation (Lima et al., 2004). The degradation or inhibition of carotenoid synthesis probably resulted in the degradation of chlorophylls in rice leaves. Many factors, such as temperature, salinity and luminosity, can cause physiological and biochemical changes in plants. These changes also occur as a result of the application of herbicides, where processes such as absorption, translocation and metabolism are directly involved in the responses observed in the species (Hess, 1985).

In work carried out by Sharma and Hall, 1991, evaluating the effect of salt stress, it was observed that salinity induces the degradation of β-carotene and a reduction in the formation of zeaxanthin, resulting in a decrease in the content of carotenoids, pigments apparently involved in protection against photoinhibition.

For the H_2O_2 variable, higher values were observed for the herbicide bispiribaque-sodium, which did not differ from the herbicide profoxidim. The control showed the highest values compared to the herbicide profoxidim, which did not differ from bispiribaque-sodium (Table 4).

Table 4. Hydrogen peroxide (H_2O_2) levels (µmol g^{-1} MF) and TBARS levels (nmol MDA g^{-1} MF) at five days after herbicide application, of rice leaves, cultivar IRGA 424 submitted to a temperature of 25°C.

FAEMUFPel, Capao Leao, RS, 2014.

Herbicides	Dose (g i.a. ha)$^{-1}$	H_2O_2 levels	TBARS levels
witness	-	29,8 a	173,8 b[1]
profoxidim	170	23,4 b	346,9 a
		32,7 a	226 2 ab
bispiribaque sodium	50	18,92	41,9
CV (%)			

[1]Lowercase letters compare herbicides using the Tukey test (p<0.05).

There was no difference between bispiribaque sodium and the control. This factor can be explained by the possible late formation of ROS, probably due to the herbicide's mode of action.

In a study carried out with rice, some herbicides were evaluated at different times after application. For penoxsulam, a herbicide that inhibits the ALS enzyme, the same mechanism of action as bispiribaque sodium, Nohatto, 2014, observed an increase in H_2O_2 production 96 hours after application.

The plants treated with the herbicide profoxidim had a lower H_2O_2 content than the control. The characteristics observed in rice, such as low absorption by the cuticle and increased metabolism, are factors that explain the crop's high selectivity to some herbicides such as butyl cyhalofop (Ruiz-Santaella & Heredia 2006). However, studies show that, in general, ACCase inhibiting herbicides have the ability to cause damage to tolerant plants (Lukatkin et al., 2013), corroborating the results of this study.

When evaluating peroxidation, there was a higher MDA content for the herbicide profoxidim, which did not differ from the herbicide bispiribaque-sodium. Reactive oxygen species (ROS) can react with lipids, proteins and pigments, causing lipid peroxidation and membrane damage (Mittler, 2002). MDA is the product of lipid peroxidation, so a higher content of this compound indicates oxidative stress (Faheed, 2012).

The increase in MDA content with the herbicide profoxidim was also noted by Langaro (2011), suggesting that the formation of phytotoxicity symptoms observed in rice crops is due to the damage caused by herbicides to biological membranes.

Studies show that irrigation shortly after the application of herbicides can make the molecule available in the root region, facilitating its absorption and favoring the occurrence of plant damage (Bond et al. 2007).

To mitigate the production of ROS, plants have developed antioxidant defense systems at the cellular level. The degree of oxidative stress in a cell is determined by the amounts of superoxide, H_2O_2 and hydroxyl radicals. Therefore, balancing the activity of the enzymes superoxide dismutase (SOD), ascorbate peroxidase (APX) and catalase (CAT) is fundamental for suppressing toxic levels of ROS in a cell (Apel & Hirt, 2004).

There was a difference between the herbicide treatments in terms of SOD enzyme activity, with activity in the control not differing from the bispiribaque sodium herbicide. The treatment with profoxidim showed lower SOD activity than the control, but this did not differ from the plants treated with the herbicide bispiribaque-sodium (Table 5).

Table 5. Activity of the enzymes superoxide dismutase SOD (μmg$^-$ 1 Protein), catalase CAT (μmol H_2O_2 min^{-1} mg^{-1} Prot.) and ascorbate peroxidase APX (μmol ASA min^{-1} mg^{-1} Prot.) five days after application of the herbicides to leaves of rice cultivar IRGA 424, subjected to a temperature of 25°C.

FAEM/UFPel, Capao do Leao, RS, 2014.

Herbicides	Dose (g i.a. ha)[-1]	SOD	CAT	APX
witness	-	46,7 a[1]	0,42 [ns2]	1,73 [ns2]
profoxidim	170	37,3 b	0,34	2,20
bispiribaque sodium	50	40.5 ab	0,38	2,10
CV (%)		15,27	20,30	24,10

[1]Lowercase letters compare herbicides using the Tukey test (p<0.05).

[2] Not significant

Among the antioxidant enzymes, SOD, found in almost all cell compartments, is responsible for catalyzing the dismutation of the superoxide radical (O_2^-) into hydrogen peroxide (H_2O_2) and oxygen (O_2), preventing the formation of the hydroxyl radical (OH.), which is a more reactive and toxic radical than the others (Gill & Tuteja, 2010). Therefore, in order to better understand the various stresses on different plant species, we are looking to correlate the levels of ROS and the activity of antioxidant enzymes in plant tissues.

Increased SOD enzyme activity was observed in maize plants treated with alachlor and metachlor (Stajner et al., 2004), glyphosate (Miteva et al., 2005), and in rice plants with the herbicide penoxsulam (Langaro et al., 2013).

Higher SOD activity in plants has been correlated with tolerance to oxidative stress (Iannelli et al., 1999). The increase in SOD enzyme activity is related to the increase in concentration and exposure time of the plant to the herbicide. The results suggest that the exposure time of the rice plants in this study may not have been sufficient to increase the activity of the SOD enzyme, a fact which is evident from the fact that the highest value was observed in the control plant. However, in a study presented by Langaro et al., 2013, the activity of the SOD enzyme with the application of penoxsulam to rice plants stabilized around 16 hours after the application of the herbicide.

For the CAT enzyme, the treatments did not differ (Table 5). These results corroborate those observed by Langaro et al., 2013. In a study assessing the activation of the antioxidant system in rice plants subjected to the application of penoxsulam, these authors observed no statistical differences in CAT activity. The results show that the lack of statistical significance in CAT activity may be due to the enzyme's lower affinity for hydrogen peroxide compared to APX.

The herbicide treatments did not differ statistically when it came to assessing the enzyme ascorbate peroxidase (APX). The plants behaved similarly to the control (no herbicide application) (Table 5).

These results corroborate those presented by Nohatto, 2014, where no difference was observed between the different herbicides applied to the rice crop. The author attributes the results to the different isoforms present in rice plants, which implies variability in the response to herbicide application. For rice, eight types of APX are described, and the results of the enzyme's activity are related to the averages between these types of enzyme, which can infer significant responses to stress.

Based on the above, it can be inferred that the use of herbicides alters the metabolism of rice. The application of bispiribaque sodium did not differ from the control in the variables analyzed, suggesting that the herbicide did not cause damage to the rice plants. However, the herbicide profoxidim is capable of causing oxidative stress, as detected by the levels of H_2O_2, total chlorophyll and carotenoids. To mitigate the effect of the herbicide, the crop activates its defense system, in particular the SOD enzyme.

Understanding the behavior of these enzymes in the face of herbicide application is an important tool for defining appropriate pesticide recommendations, with the aim of reducing the damage caused to crops and reducing the expression of maximum production potential.

Experiment 2 - Biochemical changes in irrigated rice in response to the application of seed treatment and herbicides at a temperature of 17°C.

For all the variables analyzed, there was no interaction between the seed treatment and herbicide factors. A significant difference was only observed for the herbicide factor at a temperature of 17°C.

For the total chlorophyll variable, there was a significant difference between the herbicide treatments, with an increase in chlorophyll content in plants treated with the herbicide bispiribaque-sodium (Table 6). Treatment with profoxidim did not alter the chlorophyll content compared to the control.

There are few studies in the literature on the results of determining chlorophyll levels in relation to the application of herbicides at low temperatures. It is assumed that inhibition of the synthesis of the amino acids valine, leucine and isoleucine results in altered protein synthesis, affecting the formation of some chlorophyll precursor acids (Cayon et al., 1990).

Table 6. Concentration of total chlorophyll (mg g^{-1} MF) and carotenoids (mg g^{-1} MF) five days after herbicide application, in rice leaves of the IRGA 424 cultivar, subjected to a temperature of 17°C. FAEM/UFPel, Capao do Leao, RS, 2014.

Herbicides	Dose (g i.a. ha)$^{-1}$	Total Chlorophyll	Carotenoids
witness	-	1,64 b[1]	0,13 a
profoxidim	170	1,51 b	0,07 b
bispiribaque sodium	50	2 01 a	0 12 ab
CV (%)		1088	3022

[1] Lowercase letters compare herbicides using the Tukey test (p<0.05).

According to the results, it can be inferred that the herbicide treatments induce different responses between the herbicides, stimulating the secondary metabolism of rice plants through the synthesis of antioxidant molecules, both enzymatic and non-enzymatic.

For the carotenoids variable, the same behavior was observed as for the chlorophyll variable, where the plants treated with bispiribaque-sodium had a higher content of these pigments than the control plants. The plants treated with profoxidim did not differ from the control plants or those treated with bispiribaque sodium (Table 6). An important function of carotenoids is to act as a photoprotector to prevent photooxidative damage (Cogdell, 1988, Rau, 1988). Carotenoids are able to prevent the reactive action of singlet oxygen produced by chlorophyll (Cogdell, 1988).

In a study carried out by Piesanti et al, (2013), it was shown that the chlorophyll content and photosynthetic rate did not change as a result of the use of the herbicides penoxsulam, bispiribaque-sodium and cyhalofop-butyl in the BRS Querencia cultivar, with the exception of carfentrazone-ethyl. Inhibition of chlorophyll synthesis can compromise photosynthesis, resulting in a reduction in the plant's ability to take advantage of environmental resources, such as water.

The increase in chlorophyll concentration in some herbicide treatments may be the result of chloroplast development (increase in the number of thylakoids) or an increase in the number of chloroplasts, suggesting the activation of a protective mechanism for the photosynthetic apparatus (Garcia et al., 2005).

There was no significant difference between the herbicides for the concentration of hydrogen peroxide (H_2O_2). However, there was a significant difference for the lipid peroxidation variable, where the plants treated with the herbicide profoxidim showed a higher level of TBARS when compared to the control treatment and the herbicide bispiribaque sodium, which did not differ (Table 7).

No difference was observed between the herbicides for the H_2O_2 variable. However, the accumulation of H2O2 in specific tissues and in appropriate quantities can bring some benefits to plants, mediating acclimatization and cross-tolerance to biotic and abiotic stresses (Bowler, 2000; Fluhr, 2000).

Plants from temperate regions usually show greater tolerance to cold, due to their exposure to low temperatures. This process is known as cold acclimatization, which requires numerous physiological and biochemical changes (Van Buskirk, 2006; Thomashow, 2006). However, plants continuously exposed to cold (since emergence), regardless of herbicide treatments, showed significantly equal levels of lipid peroxidation, showing that plants are able to cope with the generation of ROS, triggered by heat stress from long-term exposure. This is in addition to the effect of the IRGA 424 cultivar, which naturally confers greater tolerance to cold.

Table 7. Levels of hydrogen peroxide (H_2O_2) (μmol g^{-1} MF) and levels of lipid peroxidation (TBARS) (nmol MDA g^{-1} MF) at five days after herbicide application, in rice leaves of the IRGA 424 cultivar subjected to a temperature of 17°C. FAEM\UFPel, Capao do Leao, RS, 2014.

Herbicides	Dose (g i.a. ha)$^{-1}$	H2O2 levels	Levels of TBARS
witness	-	43,1 [ns1]	137,5 b[2]
profoxidim	170	47,8	161,5 a
	çn	ʌ 1 n	nmk
Uispuiuaque suuico			
CV (%)		16,64	309

[1] Not significant

[2] Lowercase letters compare herbicides using the Tukey test (p<0.05).

Oxidative stress corresponds to an imbalance between the rate of production of oxidizing agents and their degradation (Sies, 1991) and occurs when the production of ROS is accelerated or when the mechanisms involved in protection against ROS are impaired (Giasson et al., 2002). ROS are constantly formed in living organisms during physiological processes (Huang et al., 2007). However, in an attempt to minimize this situation, cells and their organelles (chloroplasts, mitochondria and peroxisomes) have developed an antioxidant system made up of enzymatic and non-enzymatic components.

To reduce the damage caused by oxidative stress, plants have a defense system that includes various antioxidant enzymes in different cellular compartments. Among the main enzymes is superoxide dismutase (SOD), which, together with other enzymes such as catalase (CAT) and ascorbate peroxidase (APX), promotes the elimination of reactive oxygen species (ROS), the main causes of oxidative stress (Marchezan, 2014).

Among the antioxidant enzymes is SOD, present in most cell compartments, where its main activity is to catalyze the dismutation of the superoxide radical (O_2^-) into hydrogen peroxide (H_2O_2) and oxygen (O_2), preventing the formation of the hydroxyl radical (OH.), which is a more reactive and toxic radical than the others (Gill & Tuteja, 2010). Therefore, in order to better understand the various stresses on different plant species, we are looking to correlate the levels of ROS and the activity of antioxidant enzymes in plant tissues.

In this study, there was a significant difference in SOD activity between the herbicides. The lowest level of SOD was found in the bispiribaque-sodium herbicide treatment, with no difference between the control and the profoxidim herbicide (Table 8).

Table 8. Activity of the enzymes superoxide dismutase SOD (μmg^{-1} Protein), catalase CAT (μmol

H2O2 min^{-1} mg^{-1} Prot.) and ascorbate peroxidase APX (μmol ASA min^{-1} mg^{-1} Prot.) five days after application of the herbicides to leaves of rice cultivar IRGA 424, subjected to a temperature of 25°C. FAEM/UFPel, Capao do Leao, RS, 2014.

Herbicides	Dose (g i.a. ha)$^{-1}$	SOD	CAT	APX
witness	-	30,4 a[1]	0,33 [ns2]	1,59 [ns2]
profoxidim	170	27,2 a	0,32	1,75
bispiribaque sodium	50	13,8 b	0,34	1,68
CV (%)		16,5	19,50	27,8

[1]Lowercase letters compare herbicides using the Tukey test (p<0.05).

[2] Not significant

The activity of the APX enzyme showed a significant difference between the herbicides, with its highest expression occurring in the profoxidim and bispiribaque-sodium herbicide treatments, with no difference between the two (Table 8). The increase in APX expression in plants has been demonstrated during different stress situations (Gill and Tuteja, 2010).

The APX enzyme is found in chloroplasts, cytosol, mitochondria and peroxisomes (Asada, 1999). Like CAT, APX converts H2O2 into water and o2, using ascorbate as an electron donor. However, it has different affinities for ROS. APX shows an affinity for H2O2 in the micromolar range, while CAT has a millimolar affinity.

Thus, APX seems to be responsible for the fine regulation of the response to ROS (Mittler, 2002), but it is highly important in protecting against oxidative damage in subcellular compartments where CAT is not present, such as chloroplasts where two isoforms are found, one bound to the thylakoid and the other dispersed in the stroma.The APX enzyme has a high affinity for hydrogen peroxide, while CAT has a low affinity, which makes it important in removing H2O2 from oxidative stress.

There was no significant difference in the activity of the enzyme catalase (CAT) between the herbicide treatments (Table 8). This may be due to the enzyme's lower affinity for hydrogen peroxide compared to APX.

It is proposed that changes in the activity of antioxidant enzymes exposed to herbicide application and evaluated over time are directly linked to the suppression of oxidative stress through the elimination of ROS produced by herbicide application.

Photosynthesis is one of the primary processes that can be affected by stressful situations (Chaves et al., 2009). In higher plants, the capture and storage of light energy during photosynthesis is carried out by the association of light receptor pigments and electron transport from photosystem II (PSII) to photosystem I (PSI). This process of electron transfer between PSII and PSI results in the production of ROS and is part of the plant's normal metabolism (Foyer and Noctor, 2000; Müller et al., 2001).

Based on the above, it can be inferred that seed treatment does not influence plant defense responses and that herbicide treatments induce different responses in rice plants, stimulating the plant's secondary metabolism through the synthesis of enzymatic and non-enzymatic antioxidant molecules. Situations of low temperature together with the application of the herbicide can impair absorption by the plant.

In view of the results, the lower oxidative stress of the control showed the high activity of enzymes such as SOD. Plants subjected to low temperatures may have their metabolism affected, which can lead to lower absorption and translocation of the product, a fact which explains the lower capacity of herbicides such as bispiribaque sodium to cause oxidative stress in rice plants.

Generally, plants that show a reduction in chlorophyll and carotenoid content compared to the control, for example, as well as an increase in H2O2 and TBARS content, are said to be more stressed plants. The activity of enzymes such as CAT and APX can vary according to the species, the type of stress and the time the plants were collected.

BIBLIOGRAPHICAL REFERENCES

APEL, K.; HIRT, H. Reactive Oxygen Species: Metabolism, Oxidative Stress, and Signal Transduction. Annual Review of Plant Biology, v. 55, p. 373-399, 2004.

BOND, J.A. et al, Rice cultivar response to penoxsulam. Weed Technology, v. 56, p. 961-965, 2007.

BOWLER, C.; FLUHR, R. The role of calcium and activated oxygens as signals for controlling cross-tolerance. Trends Plant Science, v.5, n.6, p.241-246, 2000.

BRADFORD, M. M. A rapid and sensitive method for the quantitation of microgram quantities of protein utilizing the principle of protein-dye binding. Analytical Biochemistry, v.72, p.248-254, 1976.

CASSOL, G.V. et al., Efficiency of herbicides in irrigated rice under intermittency. In: Brazilian Congress of Weed Science, 28. 2012. Campo Grande, MS. Proceedings. 2012, p. 192-197.

CAYON, D.G. Chlorophyll and crude protein contents in soybean (Glycine max (L.) Merril) treated with imazaquin. Revista Brasileira de Fisiologia Vegetal, v.2, p.33-40, 1990.

CHAVES, M.M.; FLEXAS, J.; PINHEIRO, C. Photosynthesis and salt stress: Regulation mechanisms from whole plant to cell. Annals of Botany, v.103, p.551-560, 2009.

COGDELL, R. The functions of pigments in chloroplasts. In: GOODWIN, T. W. (Ed.). Plant

Pigments. London: **Academic Press**, 1988. p.183-230.

CRUZ, R.P.; MILACH, S.C.K. Genetic improvement for cold tolerance in irrigated rice. **Ciência Rural**, v.30, p.909-917, 2000.

CRUZ, R.P. Genetic bases of cold tolerance in rice (Oryza sativa L.). 2001. 155f. **Thesis** (Doctorate in Phytotechnology) - Phytotechnology Postgraduate Program, Federal University of Rio Grande do Sul, Porto Alegre.

EMBRAPA - BRAZILIAN AGRICULTURAL RESEARCH Corporation. National Center for Agricultural Soil Research (Rio de Janeiro, RJ). **Brazilian soil classification system**. Brasilia: Embrapa Produçao de Informaçao, Rio de Janeiro: Embrapa solos, 2006. 412 p.

FAHEED, Fayza. Comparative effects of four herbicides on physiological aspects in Triticum sativum L. **African Journal of Ecology**, v.50, p. 29-42, 2012.

FOYER, C.H.; NOCTOR, G. Oxygen processing in photosynthesis: Regulation and signaling. **New Phytologist**, v.146, p.359-388, 2000.

GARCIA, X.V. et al. Chlorophyll accumulation is enhanced by osmotic stress in graminaceous chlorophyllic cells. **Journal of Plant Physiology**, v.162, p.650-61, 2005.

GIASSON, B.I. et al. The relationship between oxidative/nitrative stress and pathological Alzheimer's and Parkinson's diseases. **Free Radical Biology**, v.32, p.1264-1275, 2002.

GIIL, S.S.; TUTEJA, N. Reactive oxygen species and antioxidant machinery in abiotic stress tolerance in crop plants. **Plant Physiology and Biochemistry**, v. 48, p. 909-930, 2010.

GOMEZ M. A., et al. Study of the topical anti-inflammatory activity of Achillea ageratum on chronic and acute inflammation models. **Z Naturforsch** v. 54, p. 937941, 1999.

GROHS, M. Study of substances with a growth regulator effect on the physiological potential of irrigated rice. 2012. 89f. **Dissertation** (Master's Degree in Agronomy) Postgraduate Degree in Agronomy, Area of Concentration in Plant Production, Federal University of Santa Maria.

GUO, Z. et al. Differential responses of antioxidative system to chilling and drought in four rice cultivars differing in sensitivity. **Plant Physiology and Biochemistry**, v.44, p.828-836, 2006.

HEATH, R. L.; PACKER, L. Photoperoxidation in isolated chloroplasts. I. Kinetics and Stoichiometry of fatty acid peroxidation. **Archives of Biochemistry and Biophysics**, v.125, p.189-198, 1968.

HUANG, R. et al. Antioxidant activity and oxygen-scavenging system in orange pulp during fruit ripening and maturation. **Scientia Horticulturae**, v.113, p.166-172, 2007.

IANNELLI M. A. et al. Tolerance to low temperature and paraquat mediated oxidative stress in two maize genotypes. Journal of Experimental Botany, v. 50, p. 523-532, 1999.

LANGARO, A.C. et al., Production of phenolic compounds and cell damage by the application of profoxydim in rice culture. In: Congresso Brasileiro de Arroz Irrigado, 7. 2011, Balneàrio Camboriù. Proceedings of the... Itajai, 2011, p. 431-433.

LANGARO, A.C. et al., Activation of the enzymatic antioxidant system in rice plants submitted to the application of penoxsulam. In: Congresso Brasileiro de Arroz Irrigado, 8., Santa Maria, RS. Proceedings. 2013, p. 356-359.

LAW, R.D. et al. High temperature stress increases the expression of wheat leaf ribulose-1,5-bisphosphate carboylase/oxygenase activase protein. Archives of Biochemistry and Biophysics, v.386, 261-267, 2001.

LICHTENTHALER, H.K. Chlorophyll and carotenoids: pigments of photosynthetic biomembranes. In: COLOWICK, S.P.; KAPLAN, N.O. Methods in enzymology, Academic Press, 1987, p.350-382.

LIMA, A.M. Influence of light competition on root development of rice seedlings in competition with red rice and different temperatures. In: Congresso Brasileiro de Arroz Irrigado, 8., Santa Maria, RS, Anais... 2004.

LORETO, F.; VELIKOVA, V. Isoprene produced by leaves protects the photosynthetic apparatus against ozone damage, quenches ozone products, and reduces lipid peroxidation of cellular membranes. Plant Physiology, v.127, p.1781-1787, 2001.

LUKATKIN, A.S.; Treatment with the herbicide TOPIK induces oxidative stress in cereal leaves. Pesticide Biochemistry and Physiology, v. 105, p. 44-49, 2013.

MITEVA, L. et al. Alterations of the Content of Hydrogen Peroxide and Malondialdehyde and the Activity of Some Antioxidant Enzymes in the Roots and Leaves of Pea and Wheat Plants Exposed to Glyphosate. Comptes rendus de l'Acade'mie bulgare des Sciences, v. 58, p. 723-728, 2005.

MITTLER, R. Oxidative stress, antioxidants and stress tolerance. Trends in Plant Science, v.7, p.405-410, 2002.

MORSY, M.R. et al. Alteration of oxidative and carbohydrate metabolism under abiotic stress in two rice (Oryza sativa L.) genotypes contrasting in chilling tolerance. Journal of Plant Physiology, v.164, p.157-167, 2007.

MÜLLER, P.; LI, X.P.; NIYOGI, K.K. Non-photochemical quenching. A response to excess light energy. Plant Physiology, v.125, n.4, p.1558-1566, 2001.

MURATA, N.; LOS, D.A. Membrane fluidity and temperature perception. Plant Physiol, v.115, p. 875-879, 1997.

NEILSON, K. A. Proteomic analysis of temperature stress in plants. Proteomics, v.10, p.828-845, 2010.

NOHATTO, M.A. Physiological interrelations of irrigated rice with red rice and crop response to herbicides. 2014. 169f. Thesis (Doctorate in Plant Health) - Plant Health Graduate Program, Federal University of Pelotas, Pelotas.

OKUNO, K. Genetics and molecular biology research on cold tolerance of rice. In: International temperate rice conference, 3., 2003, Punta del Este. Symposia and conferences. Punta del Este: Instituto Nacional de Investigaciones Agropecuàrias, 2003.

PARK, J.E.; Selective mechanism of cyhalofop-butyl ester between rice and Echinochloa crus-galli-I. Differential response of rice and Echinochloa crus-galli to cyhalofop-butyl ester. Korean Journal of Weed Science, v. 14, p. 94-100, 1994.

PEIXOTO, P. H. P. et al. Aluminum effects on lipid peroxidation and on activies of enzymes of oxidative metabolism in sorghum. Revista Brasileira de Fisiologia Vegetal, v.11, p.137-143, 1999.

PIESANTI, S.R. et al. Physiology of rice plants submitted to herbicide application. In: Congresso Brasileiro da Ciência das Plantas Daninhas, 28, Campo Grande, MS. Proceedings... Campo Grande, 2012, p.133-137.

RAMESH, K. et al. Chlorophyll dynamics in rice (Oryza sativa) before and after flowering based on SPAD (chlorophyll) meter monitoring and its relation with grain yield. Journal of Agronomy and Crop Science, v.188, p.102-105, 2002.

RAU, W. Functions of carotenoids other than in photosynthesis. In: GOODWIN, T. W. (Ed.). Plant pigments. London: Academic Press, 1988, p.231-255.

RUIZ-SANTAELLA, J.P.; HEREDIA, A. Basis of selectivity of cyhalofop-butyl in Oriza sativa L. Planta, v. 223, p. 191-199, 2006

SHARMA, P.K.; HALL, D.O. Interaction of salt stress and photoinhibition on photosynthesis in barley and sorghun.Journal of Plant Physiology, Stuttgart, v.138, p. 614-619, 1991.

SOARES, A.M.S.; MACHADO, O.L.T. Plant defense: Chemical signaling and reactive oxygen species. Revista Tròpica - Ciências Agràrias e Biológicas, v.1, p.1-19, 2007.

SONG, N. H. et al. Biological responses of wheat (Triticum aestivum) plants to the herbicide chlorotoluron in soils. Chemosphere, v. 68, p. 1779-1787, 2007.

STAJNER, D. et al. Herbicide Induced Oxidative Stress in Lettuce, Beans, Pea Seeds and Leaves. Biologia Plantarum, v. 47, p. 575-579, 2004.

STEINMETZ, S.; INFELD, J.A.; MALUF, J.R.T.; SOUZA, P.R. de; BUENO, A.C. Agro-climatic zoning of irrigated rice in the state of Rio Grande do Sul: recommendation of sowing times by municipality. Pelotas: EMBRAPA-CPACT, 1996. 30p. (EMBRAPA-CPACT. Documentos, 19)

TAIZ, L.; ZEIGER, E. Plant physiology. 4.ed. Porto Alegre: Artmed, 2009. 819p.

VAN BUSKIRK, H.A.; THOMASHOW, M.F. Arabidopsis transcription factors regulating cold acclimation. Physiol Plant, v.126, p. 72-80, 2006.

YOSHIDA, S. Fundamentals of rice crop science. Los Banos: IRRI, 1981. 277.

Buy your books fast and straightforward online - at one of world's fastest growing online book stores! Environmentally sound due to Print-on-Demand technologies.

Buy your books online at
www.morebooks.shop

Kaufen Sie Ihre Bücher schnell und unkompliziert online – auf einer der am schnellsten wachsenden Buchhandelsplattformen weltweit! Dank Print-On-Demand umwelt- und ressourcenschonend produziert.

Bücher schneller online kaufen
www.morebooks.shop